UNIVERSITY OF MAINE
LIBRARIES

RAYMOND H. FOGLER LIBRARY
GIFT OF
Prof. J.J. Bhatt

Introduction to

GEOCHEMISTRY and GEOLOGY of SOUTH WALES' MAIN LIMESTONES

(LOWER CARBONIFEROUS)

J. J. Bhatt, Ph.D.

MODERN PRESS, INC.

Copyright © 1976 by J. J. Bhatt

All rights reserved

This book, or parts thereof, must not be used or reporduced in any manner without written permission of the author.

Printed in the United States of America

Library of Congress Catalog Card No. 76-23486

ISBN No. 0-9600900-1-0

Modern Press, Inc.
Cranston, Rhode Island

THROUGH THIS VENTURE
I RECKON
ORDER IN NATURE. . . . . .

PREFACE

The South Wales Main Limestones of Lower Carboniferous age constitutes one of the classical stratigraphic segments of the British geology. Moreover, these carbonate rocks have provided nearly for a century, a complex but challenging task for stratigraphic, structural and economic explorations.

I have been prompted by two principal reasons to write this book. First, the unusual and surprising response and interests my recent publications dealing with marine carbonates of South Wales (U.K.) and Nevada have received in recent years — quite happily on a global level. And second, my personal objective to provide an updated and comprehensive account of the Main Limestone Series which will serve as a research guide to students of carbonate geology particularly interested in undertaking similar work in Gt. Britain and elsewhere.

In this book I have made an attempt to explain the process of *dolomitization* — an important aspect of diagenesis of the marine carbonates by considering the Main Limestones of South Wales as a case study. My proposed hypothesis explaining the origin and dolomitization of the Main Limestones of South Wales is based upon vigrous geological, petrological and chemical analyses of over six hundred samples collected from fifty key localities and careful field geologic observations of the study area carried out about a period of two years during my studentship years in Gt. Britain. However, the actual synthesis of my work, consequently formulation of the hypothesis surfaced after three more years of rather an obsessed meditation of the problem under study in the United States. In between the process of gradual refinement of my thoughts, that is of the hypothesis under question — I had submitted bulk of the work as my Ph.D. thesis. It is only in the subsequent years primarily through the channel of publications, continuous correspondence through letters, and discussion of the subject at various geological conferences with collegues and other such related sources that final finishing touches of my work began to take place.

Several people have helped me in realizing my educational and professional goals. In particular, I am indebted to Professor J. G. C. Anderson, Head, Department of Geology at the University of Wales, Cardiff for giving me an opportunity to investigate the Main Limestone Series as my doctoral

research top
in subseque
help and a
am grateful
T. N. Georg
University, (
outstanding
South Wales
of my work.
 I am a
the Univers
Botany and
Centre and 1
 I am inc
of Wales, C
grateful to
Managemen
his help ar
provided me
 During
warm hearte
and certain
cherish tha
adventure c
regions. Thi
dimension
far beyond
I believe th
tion and dip
 I owe
channelling
joie de vivre.

ILLUSTRATIONS

			Page
TABLE I	Stratigraphic Column of the Main Limestone Series in South Wales		22
II	Common Algal Structures in the Main Limestone Series		42
III	Ca/Mg Ratio in the Main Limestone Series of South Wales		50
IV	Magnesium Content in the Main Limestones of the Taffs Well Region		52
V	Geochemical Data on the Main Limestone Rocks in the Type Section at Tongwynlais, near Cardiff		54
VI	Ti/Al Ratio Values in Selected Samples of the Main Limestones		61
VII	Trace Elemental Distribution in the Main Limestone Series		62
FIGURE 1	Index Map of the Study Area		2
2	Regional Variation in Thickness of the Main Limestone Series		16
3	Stratigraphic Section of the Main Limestone Series at Tongwynlais		26
4	Geological Map of the Tongwynlais Area		28
5	Regional Distribution of Microfacies in the Main Limestone Series		47
6	Bimodal Distribution of the Mg Content in the Main Limestone Series		52
7	K/Na and Ca/Mg Ratios Relation in the Main Limestone Series		56
8	Fe/Mn Behavior in the Main Limestone Series		57

ILLUSTRATIONS (continued)

			Page
FIGURE	9	Trace Elemental Distribution in the Main Limestone Series...............	59
	10	History of Deposition during Stages A and B.......................	66
	11	Regional Pattern of Chemical and Petrographic Parameters	74
	12	Depositional Model of the Main Limestone	80
PLATE I (a-f)		Field geologic characteristics of the Main Limestone Series................	18
	II	Common minerals seen under thin section	39
	III	Dolomitized crinoidal limestone...........	41
	IV (a-f)	Major microfacies.......................	44
	V	Stromatolitic limestone...................	76
	VI	Evaporite minerals inferred from the petrographic evidence.................	78
	VII	Major types of dolostone................	86

Chapter One
INTRODUCTION

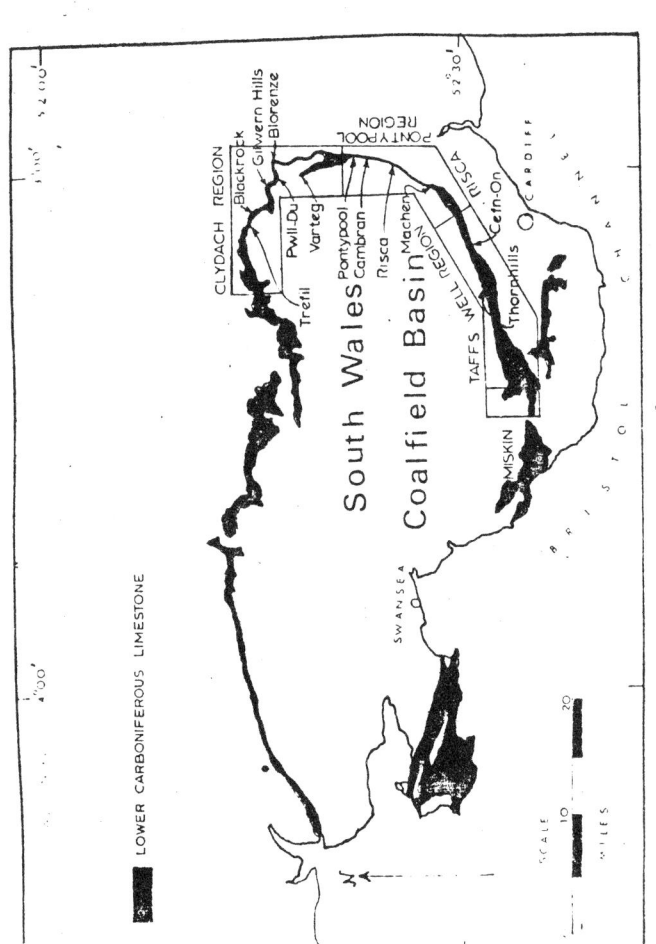

Figure 1. Index map showing the location of the study area.

CHAPTER ONE

INTRODUCTION

The Dolomite Problem:

The origin of dolomite or dolostone[1] constitutes the so called "dolomite problem" and at present remains one of the unresolved topics in sedimentary carbonate geology. The crux of the "dolomite problem" remains in the puzzling fact that although it (dolomite) is so abundant under natural conditions, still no equivocal synthetic dolomite has been successfully produced under the normal laboratory conditions. This particular stated problem has been approached by various workers following either the experimental or the geological methods.

The experimental method began more than a century ago (Von Morlot, 1947 in Fairbridge, 1957, also Seigal, 1961) but did not become quite significant until the advent of the sophisticated X-ray technique in the field of carbonate geochemistry (Graf and Goldsmith, 1965; Goldsmith and Graf, 1958; Chave, 1952). In recent years, the experimental approach to the problem of dolomite genesis has also been supplemented by thermodynamics and solubility studies (Garrels et al, 1960; Goldsmith, 1959; Kramer, 1959; Halla et al, 1962; Berner, 1966; Sass, 1965, Hsu, 1967; Udowski, 1969, and others), bacteriological studies (Oppenheimer and Master, 1963; Carroll and Greenfield, 1963; Nehrer and Rohrer, 1958) and isotopic studies (Daughtry et al, 1962; Degens and Epstein, 1964). For example, the isotopic studies of some workers (notably among them Degens and Epsteln, 1964) showed that all dolostones since the Cambrian are the diagenetic products of metasomatism of $CaCO_3$ by $MgCO_3$ occurring in solid solution.

The beginning of the geological method was largely based upon the qualitative geological field and certain petrographical criteria for recognizing the origin and time of alteration of dolostones (Dixon, 1907; Van Tuyl, 1917; Steidman, 1926). However, this method did not become more significant either until after the worldwide intensive studies of the Recent

---

[1]In this study, dolomite implies the mineral or crystal of dolomite itself whereas dolostone means rock composed of dolomite mineral. The term dolostone is used in the manner described by Friedman and Sanders. (1967).

corbonate sediments (Alderman and Skinner, 1957; Alderman and Von der Borch, 1960, 1961; Curtis et al, 1963; Sherman, 1963; Kinsman, 1964; Deffeyes et al, 1965; Shinn et al, 1965; Illing et al, 1965). The geological method is based upon the assumption that the processes which are operating today in the Recent sediments also operated in the formation of the carbonate rocks of the geological past. However, as Fairbridge (1964) observed, the modern analogies are not completely perfect and it may be that not all dolomite rocks have originated in the same way as those found in the Recent sediments.

In general four categories of sedimentary dolomitization in relation to the primary environment of formation are recognized in the geological column. They are as follows:

(1) Syngenetic dolostone
(2) Early diagenetic dolostone
(3) Late diagenetic dolostone, and
(4) Epigenetic dolostone.

The syngenetic dolostone implies direct chemical precipitation from sea water (*Primary dolomite* of Rodger, 1954) unless otherwise specified.[2] The early diagenetic dolostone refers to the metasomatism of $CaCO_3$ by $MgCO_3$ occurring shortly after deposition of the original calcareous sediments, whereas, the products of the later dated metasomatism of $CaCO_3$ by $MgCO_3$ are termed the late-diagenetic dolostone. Finally, the epigenetic dolostone (Friedman and Sander, 1967) indicates the action of secondary Mg-rich solution moving through the post-depositionally fractured rocks.

At present, the experimental and the geological observations of the Recent and ancient sediments overwhelmingly favor the early diagenetic origin for most dolostones. The syngenetic dolostones were common in the Pre-Cambrian and some early Paleozoic times but since then their large scale deposits are absent (Fairbridge, 1967).

The Dolomite Problem in South Wales:

The South Wales dolostones, which consititute the main subject matter of this study belong neither to the Pre-Cambrian nor to the early Paleozoic Groups, but are largely confined to the Lower Carboniferous (approximately Mississippian) Main Limestone Series comprising the ZC_1, C_2S_1, and the basal S_2 zones (Vaughan, 1905)[3]. Geographically, the dolostone is present predominantly between the Taffs Well region and the Clydach region fringing the south, east, and the northeast rims of the South Wales Coalfield Basin (Fig.1). The transition in lithofacies from limestone into dolostone both vertically and laterally begins near the Miskin area and becomes dominantly dolostone as the Main Limestone rocks approach the Taffs Well region in the east.

The present views on the origin of dolostones of the Main Limestone Series are divided into two schools of thought. One favors a "contemporaneous" origin (Dixon, 1907; Dixon and Vaughn, 1912; Dixey and Sibly, 1918) whereas the other supports their diagenetic or "subsequent" origin (George, 1954; 1956a). Dixon (1907) was the chief proponent of the "contemporaneous" origin hypothesis, and used largely geologic field and some petrographic criteria such as iron staining, the relative size of dolomite rhombs, preference of dolomitization for ooliths and corals, and the lateral extent of the dolomitized sediments. George (*op. cit.*) on the other hand, based his conclusion upon his field studies which concentrated along the northern rim in Breconshire, later extending into the area between Blorenge and Risca. One of his major arguments against the "contemporaneous" origin of dolomite was the inapplicability of Dixon's (1907) criteria to the corresponding rocks of his area.

In recent years, Bhatt (1976) recognized presence of major types of dolostones from the Main Limestone Series. They include Syngenetic, Early-diagenetic, Late-diagenetic, and Epigenetic. Bhatt (*ibid*) concluded that Early-diagenetic and the Late-diagenetic types were common in South Wales. Major points of his published work are incorporated in this book.

---

[2]See ch. 9 for further details.

---

[3]Subsequently applied in the area between Risca and Miskin by Dixey and Sibly (1918).

ZC_1 = *Zaphrentis* − *Caninia* zone, C_2S_1 = Upper *Caninia* − Lower *Seminula* zone, and S_2 = Upper *Seminula* zone (see Table I).

Chapter Two
REVIEW

CHAPTER TWO

PREVIOUS WORK

Geological contributions by principal workers on the Main Limestone Series (Lower Carboniferous) in South Wales is briefly summarized as follows:

Dixon's Contribution:

The major contribution to the study of dolomitization in South Wales came from Dixon (1907) who developed a hypothesis of the "contemporaneous" origin of dolomite.[4] Since his hypothesis influenced the minds of the later workers in South Wales, it is pertinent to clarify at the outset Dixon's usage of the term "contemporaneous" origin. The context of his definition (Dixon, 1907) which was also quoted by Dixey and Sibly (1918, p. 117) is as follows:

> From the fact that dolomite is comparable with that of Gower, and persists as such for upwards of 15 miles, it may be inferred that much of the dolomitisation was "contemporaneous" rather than the result of subsequent alteration along fault veins and faults. That the rocks were first calcitic is shown by their containing crinoidal remains, and the expression "contemporaneous" is intended as implying dolomitisation while they were still under the influence of the Carboniferous Limestone sea.

Thus, Dixon's "contemporaneous" dolomite suggests in essence either an "early diagenetic" or "penecontemporaneous" origin rather than "contemporaneous" or "syngenetic" in *sensu stricto*.

The subsequent geological study of the Gower region (Dixon and Vaughan, 1912) further supported Dixon's postulate of "contemporaneous" dolomite. These workers contended that the dolomite was "contemporaneous" in origin and that the limestone was deposited as unstable calcite or aragonite which changed into dolomite soon after their deposition.

Dixey and Sibly's Contribution:

Dixey and Sibly (1918) studied the Lower Carboniferous Limestone between Miskin and Risca applying Vaughan's

---

[4]Earlier views on dolomite were expressed by some workers (for example, Strahan, 1901) but Dixon's (1907) work marks the beginning of a firmer recognition of the problem in South Wales.

(1905) zonal classification. They showed that the Main Limestone Series in the area had undergone a major change of lithologic facies as the strata of undolomitized limestone with only subordinate dolomite present in the west (Miskin area) gave way to an unbroken sequence of dolomite in the east (*i.e.* in Taffs Well). According to them, such rapid changes were due to gradual increase in vertical extent of primary dolomitization. Furthermore, they noted that the fauna in such rocks although obscured by dolomitization did maintain a standard facies. Dixey and Sibley (*op. cit.*) also showed that in the east of the Taffs Well area, the lower beds (ZC_1) maintained the characters of crystalline dolomite with the remnants of the standard facies but the upper beds (C_2S_1) had evolved into a "*Modiola* phase" of considerable thickness, that is, a phase of calcitic beds. They (Dixey and Sibly, 1918) concluded that "contemporaneous" dolomite was predominantly present in the Taffs Well region and it was formed in the manner envisaged by Dixon (1907), whereas, the "vein-dolomite" was formed much later by metasomatism of calcite into dolomite due to the action of Mg-bearing solution circulating through the fault zones and through more diffuse permeation. Finally, according to these workers, the limestone which had been partly altered by "Contemporaneous" dolomitization could be further affected by such vein dolomitization, but at quite a later date.

Robertson and George's Contribution:

Robertson (1927) and later Robertson and George (1929) investigated the area around the Clydach region. Robertson (*op. cit.*), using his field observations and chemical analysis of a very few selected samples of the Main Limestone demonstrated a systematic but an abrupt lateral change in dolomitization in the area. He showed that the change from the unaltered oolitic limestone into crystalline dolomite was complete when it had 30% calcium carbonate replaced by magnesium carbonate. Robertson (*op. cit.*) observed in one limestone sample with its laterally opposite ends consisting of two different carbonate facies, one an unaltered oolitic limestone and the other crystalline dolostone; to him, this illustrated a more rapid lateral than vertical change. Also, in Llanelli quarry, Robertson (*op. cit.*) observed that limestone lying immediately below the Millstone Grit was dolomitized but that the degree of dolomitization decreased downwards. From such geological observations, he concluded that the dolomitization in this area was due to the subsequent processes caused by the "Vertical communication" of Mg-rich solution from fault channels or by an immediate superposition between the limestone and Millstone Grit.

George's Contribution:

George (1954, 1956a, 1956b) investigated the areas in Breconshire and the Blorenge to Risca encompassing the northwest and east rims of the South Wales Coalfield Basin. Using his field observations, including the abrupt changes of lithology from unaltered oolitic limestone into granular dolomite, patchy development of dolomitization and the failure of Dixon's criteria, George (*op. cit.*) concluded in favor of "subsequent" origin. He argued that such scattered rocks could scarcely have been formed in a uniform environment by direct precipitation between the depositing sediments and immediately overlying sea water. Furthermore, he attributed their origin to magnesium rich waters percolating through a thickness of sediment have non-uniform permeability. George (1969) has synthesized and summarized the geological aspects of the Main Limestone Series.

Bhatt's Contribution:

In recent years, Bhatt updated geochemistry and petrology of the Main Limestone Series in South Wales. His work involved in explaining the origin and diagenesis of these rocks including the process of dolomitization, discovery of evaporite deposits, and general classification (1971, 1972, 1973, 1974, 1975, and 1976).

The economic significance of the Lower Carboniferous Limestone has been treated in detail as a part of the general mineral wealth of South Wales by Anderson (1960, 1971). In recent years, regional geological accounts of South Wales providing relevant information on the Main Limestone Series have been presented by Owen (1964), Squirrell *et al* (1969), and George (1958, 1970).

Summary:

The evolution of geological research of the Main Limestones in terms of major contribution by various workers may be summarized into three phases. The *first phase* initiated near the turn of the century. During this phase, bulk of the geological investigation so done was largely conventional. However, two students of the Main Limestone Series deserve mention here. Dixon (1907) who so skillfully recognized the existence of his "contemporaneous" dolomite at a time when carbonate geology was in its embryoic stage. The second man was Vaughan (1911) who successfully established the paleontologic classification of the Lower Carboniferous formation including the Main Limestone Series. The *second phase* began more effectively in the middle of the century when George (1954, 1956a) very creditably deciphered and re-established the stratigraphic column of these rocks in the Clydach region of this study. Also, works of Blundell (1952) and T. R. Owens (1954, 1964) during this phase are noteworthy. The *third phase* began in early seventies when this author undertook his geochemical and petrological examinations of the Main Limestone Series.

Chapter Three
GEOLOGIC SETTING

CHAPTER THREE

GEOLOGICAL SETTING

Introductory Remarks

The Lower Carboniferous carbonate rocks constitute one of the most prominent geological outcrops in South Wales in terms of thickness, lithology, and economy. The Main Limestone Series of the Upper Tournasian and Lower Visean is characterized by the presence of dolomitized limestone which becomes progressively more complex in terms of their alteration (dolomitization) through the area studied.

Location

The Main Limestone Series outcrop is a narrow tract less than one mile wide and stretches around the southern rim of the Coalfield Basin for over twenty-five miles from Miskin through Taffs Well area turning northward between Risca and Pontypool and later northeastward in the Clydach region (Fig. 1). In general, the area of investigation embraces part of Glamorganshire, Monmouthshire, and Breconshire counties. Throughout the text of this book, "Taffs Well region" will be used as a comprehensive term to denote Taffs Well proper, Thornhill, Cefn-On, Draethen, Machen and the southern part of Risca (in part); and "Clydach Region" will include Clydach area proper, Blackrock, Gilwern Hills, Pwll-du, Varteg, Abersychan, Pontypool (in part). The "Risca-Pontypool region" will include Cwn-ynisycoy, Cwmbran, Ysgbour-newdd, Risca (northern part). Finally, "Miskin Area" will indicate the transitional zone where the limestone changes into dolostone towards the Taffs Well area.

Physiography

The Carboniferous outcrop in the study area is dissected by series of rivers, namely, from southwest to northeast, the Ewenny, Taff, Rhymney, Ebbw Vale, Afon-Lwydd, and Clydach. Most of these rivers are antecedent flowing without regard to geological structures and consequently their traverses across the Carboniferous outcrop have enabled them to form success-

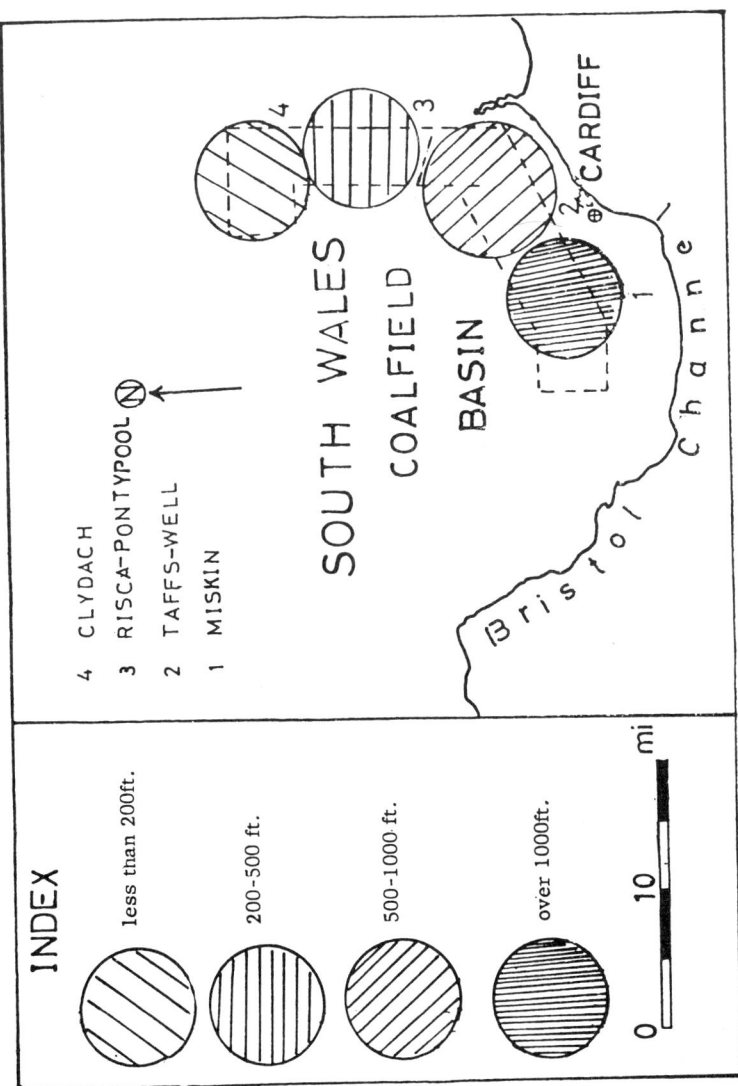

Figure 2. The regional variation in thickness of Lower Carboniferous carbonate rocks in the study area. The abrupt changes in thickness near Risca are shown by stripled lines.

fully the Main Limestone escarpments throughout the area of study. The Taffs Well gorge is a notable example (Plate Ia). The rivers Taff and Ely, however, cross the Carboniferous Limestone on lines of dip faulting (Dixey and Sibly, 1918, p. 115).

In Miskin and the Taffs Well region, the relief is relatively low ranging between 100-200 feet above sea level except for some local higher altitudes, for example, in Garthwood, where the height is slightly above 1000 feet. The general topography increases northward near Risca and Pontypool where it reaches between 500-1000 feet but farther north and northeastward it gradually rises in the Clydach region to 2000 feet and above. In general, the Clydach region has a relatively much more complex terrain and a far higher relief than those of the Miskin and the Taffs Well regions (Fig. 2).

General Structures and Tectonics

The major tectonic elements in South Wales were largely moulded by the *Armorican orogeny* and to lesser extent by the *pre-Armorican* and *Alpine* disturbances (Anderson, 1960; Anderson and Owen, 1967). The Coalfield Basin with its characteristic eastwest axis is chiefly due to the *Armorican* movement and causes a serrated outcrop of the Lower Carboniferous rocks. The significant local structure includes the Castell Coch anticline and the Tongwynlais syncline in the Taffs Well area. These local structures expose a good section of the Main Limestone Series. In addition, numerous small scale faults, folds, and veins are present in the Taffs Well region. The Taffs Well, Cefn-On and Traventhin faults are noteworthy. In general, local faulting in the Main Limestone rocks is confined to dip-faults which relate to the north-northwest fault system of the coalfield (Dixey and Sibly, 1918).

Plate I.

Plate I. Field Geologic observations of the Main Limestones.

    (a) Main Limestone Series exposed in the Garth Hills facing the Taft River near Cardiff City.

    (b) Alternate deposition of the shale (thin) and dolostone (thick) in a section at Tongwynalais, near Cardiff.

    (c) Secondary vug observed in the Main Limestone Series near Cardiff.

    (d) Main Limestone Series of *Upper Semirula* (S_2 zone) seen near the Portobello quarry in Tongwynalais, near Cardiff.

    (e) Nodular chert (white) observed in the Calcite-Mudstone Group in the Varteg Hills, Clydach region.

    (f) Alternate deposition of the oolitic limestones and dolostones seen in the Black Rock Quarry, Clydach.

Chapter Four
STRATIGRAPHY

AMERICAN EQUIVALENT	EUROPEAN EQUIVALENT	LITHOLOGIC CLASSIF.	ZONES	TAFFS-WELL REGION (DIXEY & SIBLEY, 1918)	CLYDACH REGION (GEORGE, 1954, 1956)	
MISSISSIPPIAN	Visean	MAIN LIMESTONE SERIES	S2	Main Seminula Zone	Main Seminula Zone	
			C2S1	Lower Seminula and Upper Caninia	CALCITE-MUDSTONE GROUP	LITHOLOGIC DIVISION
			ZC1	Lower Caninia and Zaphrentis zone	OOLITE Group / Zaphrendian Group	
	Tournasian	LOWER LIMESTONE SHALE SERIES	K Zone	Cleistopora zone	Cleistopora zone	
DEVONIAN OLD RED SANDSTONES						

Table I. Stratigraphic classification of Lower Carboniferous Main Limestones in South Wales.

CHAPTER FOUR

STRATIGRAPHY

Stratigraphy of Main Limestone Series

The term "Main Limestone" stems from Strahan's (1909) two-fold lithological classification which was designed for mapping purpose. According to his scheme, the lower division was the Lower Limestone Shale (roughly corresponding to K zone) and the upper division constituted the Main Limestone Series (comprising ZC_1, C_2S_1, and the S_2 zones of the Vaughan classification, (Vaughan, 1905).[5]

The Lower Limestone Shale Series is abruptly overlain by the Main Limestone Series. However, the delineation of the boundary between the ZC_1 and C_2S_1 zones within the Main Limestone Series by faunal criteria becomes difficult or impossible where the dolomitization has obliterated almost all the fossil impressions and the useful structures as for example in the Taffs Well region (Dixey and Sibly, 1918). The Upper *Seminula* zone (S_2 zone) is easily distinguished from ZC_1 and C_2S_1 zones below by its lithologies which in the S_2 zone are mainly calcitic limestone with incomplete dolomitization (from slight to considerable). In the Taffs Well region, the upper beds of the Series are characterized by the presence of "Vein dolomitization" belt (Dixey and Sibly, 1918). The stratigraphic classification based upon Vaughan's paleontological scheme was more successfully applied to the Lower Carboniferous rocks between the Miskin and Risca areas by Dixey and Sibly (1918) than in the Clydach region. George (1954, 1956a), on the other hand, reestablished the stratigraphy of the Main Limestone Series in Breconshire (including the Clydach region) by developing the lithofacies concept; he proposed two divisions, (1) a *Zaphredian* or the Oolite Group below and (2) a Calcite-mudstone Group above. Thus, his classification has helped circumvent the short comings of the paleontological classification which was adopted earlier by Robertson (1927).[6] The general usage of stratigraphic nomenclature and classification of the Lower Carboniferous System is presented in tabular form (Table I).

---

[5] This type of classification was applied earlier in the Taffs Well region by Dixey and Sibley (1918).

[6] Also adopted by Robertson and George (1929).

Lithology

The Main Limestone Series contains predominantly limestones (in varying degrees of alteration) and the dolostones. Significant vertical and lateral changes in the lithological characters in term of alteration from limestone into dolostone are observed especially eastward from the Miskin area towards the Taffs Well area. Also, abrupt changes in the lithological characters commonly occur within distance of a few yards in the Clydach region where the unaltered oolitic limestone changes abruptly into a granular dolostone (George, 1954, and 1956a). In general, in the Taffs Well region, the Main Limestone beds are characterized by the presence of heavily iron-stained coarse to medium-grained highly dolostone associated with frequently occurring secondary calcite veins, some occasional calcite vugs, accessory minerals like calcite bartes and gelena and by the local concentration of iron ores (for example, in the Garthwood area). In the Miskin area, fossils show relatively better development (or preservation) especially in the lower part of the Series where these rocks have been saved from intense dolomitization. I addition, this area is characterized by relatively less iron-staining of the rocks.

Distribution and Thickness

The Main Limestone outcrop is confined within a narrow tract from Miskin through Taffs Well, Risca, and Pontypool to the Clydach region. Between Risca and Pontypool, the Main Limestone is characterizer by the presence of the Lower ZC_1 zone only and in the vicinity of Blorenge and Myndd-Garn-Fawr, the Series disappears (George, 1956a); this is due to the gradual disappearance of younger beds between Taffs Well and Pontypool, and farther north to conspicuous overstepping by the Millstone Grit. Even at Machen, the *Seminula* or the S_2 zone completely disappears.

In general, overstepping northwards and eastwards of the Main Limestone rocks by the Mill stone Grit has been ascribed to tectonic activity and shallow water sedimentation (Dixey and Sibly, 1918; Cox 1920).

The thickness of the Series under study varies considerably from the southwest at Miskin to the northwest in the Clydach region. For example, the total thickness of the Main Limestone and the Lower Limestone Shale in Miskin is estimated by

Dixey and Sibly (1918) as 2750 feet; in Taffs Well proper it is 1750 feet, in Thornhill it is 1000 feet, in Machen it decreases to 920 feet, and near Risca it is 800 feet. Half a mile north of Risca in Ysgbour-newydd, the thickness of the series falls to 150 feet but two miles farther north in Pontypool it is around 200 feet. However, near Clydach it falls to less than 200 feet which gradually dwindles to a few feet or none in the extreme northern part (Fig. 2). George (1956a) determined the thinning of the Lower Carboniferous limestone in relation to the Millstone Grit Series overstep between Risca and Blorenge and according to him, the thinning was 1 to 60 in the north, and 1 in 12 in Risca. The dip of the beds at the contact of the Lower Limestone Shale and the upper Main Limestone vary geographically. For example, there is generally a low dip in the Miskin area, increasing slightly through the Taffs Well region (22° to 45°) to 70° at Thornhill, then falling steadily between 40° and 60° between Machen and Risca and in its neighboring localities. Near Cwmbran the Main Limestone beds are vertical, but farther north the dip tends to be lower, for example, 23° in Cwm Ynisycoy quarry and 12° at Traventhin.

Field Notes on the Main Limestone Series

The geological field notes on the Main Limestone Series are discussed in sequence from Tongwynlais (which marks the chosen type section of this study), the Taffs Well area in general, Cefn-On, Rudry, Draethen, Machen, Risca, Pontypool, Abersychan, Varteg, and the Clydach region (Fig. 1).

TYPE SECTION AT TONGWYNLAIS: Tongwynlais is located about five miles northwest of Cardiff (Fig. 1), where the thickest and almost complete development of the dolomitized Main Limestone Series is attained. The almost complete exposure of these rocks has been made possible on account of the local widening of the outcrop by the Castell-Coch anticline and the Tongwynlais syncline. Moreover, the erosive action of the Taff River has brought out their exposure in the form of an excarpment (Pl. la). Tongwynlais was chosen as the type area because in addition to their natural exposure in the gorge. Construction of a new road gave almost continuous exposure and provided fresh speciments especially in beds otherwise normally not exposed.

In the type section, the contact between the Carboniferous Lower Limestone Shale (K zone) and the Upper Main Limestone

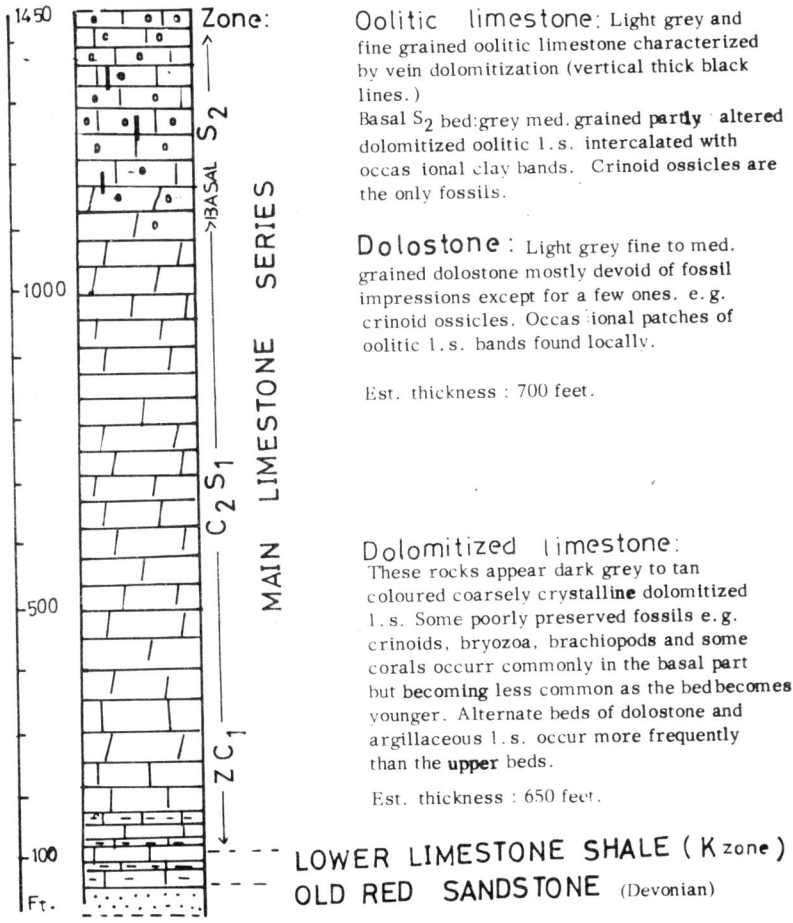

Figure 3. Stratigraphic section of the Lower Carboniferous Main Limestone rocks at Tongwynlais, near Cardiff.

is abrupt. The lower most part of the Main Limestone contains slightly altered grey to dark-grey crinoidal limestone. These rocks are coarse grained to medium grained and contains crinoids, brachiopods, some corals, cryozon, and foraminifers. The general state of fossil preservation is rather poor. These rocks frequently alternate with argillaceous limestone bands (shale-partings) of less than an inch to three feet in thickness (pl. lb). Iron-staining and calcite veins are quite common especially towards the younger beds. The thickness of this basel bed (ZC_1) is estimated as 650 feet.

Traversing toward the old Walnut Tree viaduct, these rocks (as mentioned above) give way to light-grey, fine-medium grained dolostones characterized by the extremely poor preservation of fossils; for example, crinoidal ossicles are the only surviving fossil impressions). These beds were considered as the C_2S_1 zone on the basis of changes in lithological characters. However, no definite delineation of the boundary between ZC_1 and C_2S_1 zones is possible in light of the complete obliteration of guide fossils (see also, Dixey and Sibly, 1918). The thickness of this bed is estimated as 700 feet. The degree of iron-staining is prominent in this bed and so is the development of secondary calcite veins. A local calcite vug formation is present in this bed (Pl. lc).

Approximately 250 yards north of the Walnut Tree bridge, the C_2S_1 zone passes into a relatively less dolomitized, fine grained oolitic limestone, which presumbably forms the basal beds of the *Seminula* or the S_2 zone (Dixey and Sibly, 1918). Further north, the oolitic limestone with the characteristic development of "vein-dolomitization" is encountered in the Portobello quarry (Pl. ld). Their thickness is estimated as 150 feet. The Main Limestone at the top is capped by the recent alluvial deposits.

TAFFS WELL AREA IN GENERAL: In addition to their exposure in the Tongwynlais area along the Motor Road, the Main Limestone Series occur at several other places in the vicinity, notably at the Garth Wood and the Forest Fawr areas. In Garth Wood, the lower part of the Main Limestone is found in a disused quarry (12208210) along the Morganstown road, where about 100 feet of coarse grained, hard, compact, ant crystalline massively bedded tan dolostone is exposed. In the Steetly quarry (12408279) about 100 feet of higher beds containing yellowish brown, medium to coarse grained and crystalline dolostone is exposed. These rocks are heavily

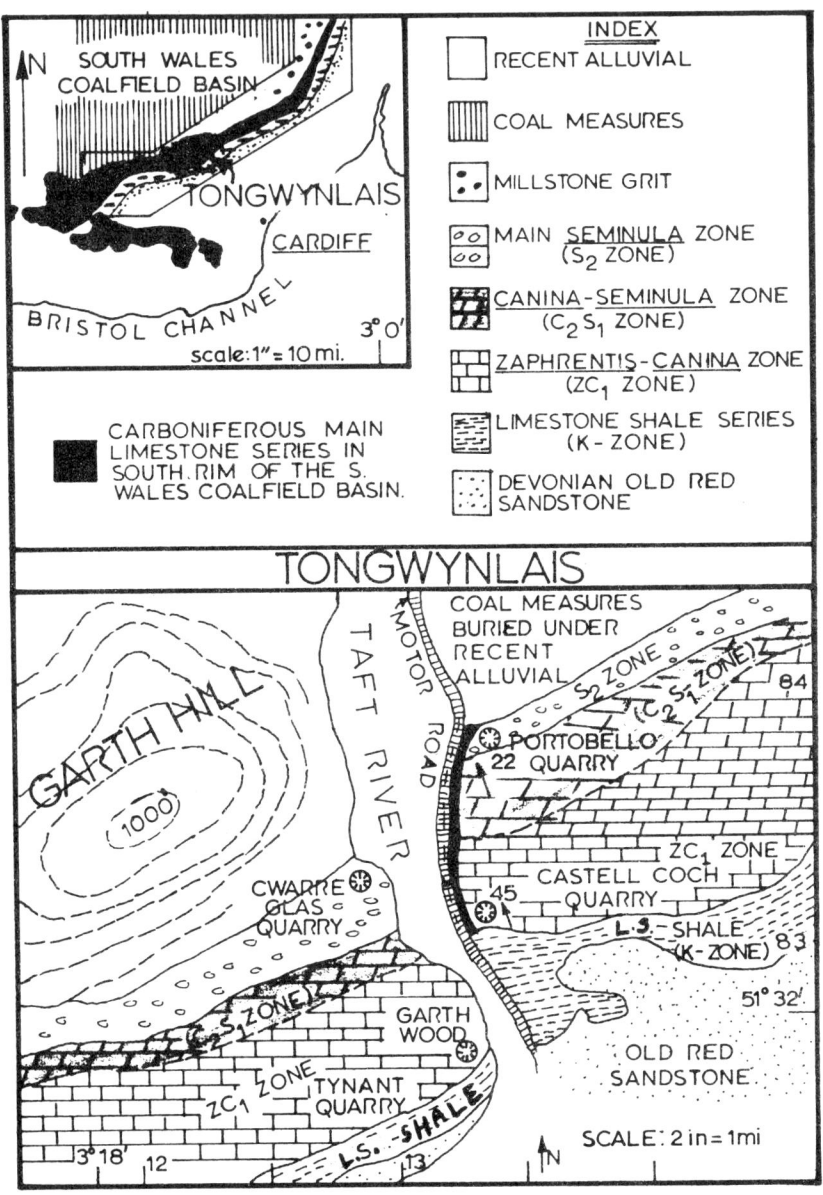

Figure 4. Location of the Main Limestone section in Tongwynlais, Cardiff District. Thick line (black in lower figure) is the studied section of this report.

iron-stained and are characterized by the development of secondary calcite veins generally running perpendicular to the bed. The upper part of the Main Limestone (S_2 zone) is found in the Gwarre Glas quarry (12208290). These rocks are mainly oolitic limestone and are characterized by the "vein dolomitization" similar to those found in the Portobello quarry.

In Garth Wood proper, the Main Limestone Series is characteristically hematized on a local scale. The iron ore, although of an economic significance in the past, is present only in the form of scattered chambers. The "Slide-pit" chamber in the Forest Fawr is noteworthy in this respect. More exposures of the Main Limestone are found in Gelli quarry (14498324) and Bwlcy-y-cwm quarry (14508404), but apart from minor local variation, the lithological characters are similar in these areas.

Towards the eastern margin of the Forest Fawr, the Main Limestone is found in local disuses quarries near Thornhill, namely, Blaen-Nofydd quarry (16058465) and a smaller quarry (15828468) located along the A470 road between Cardiff and Caerphilly. In the Blaen-Nofydd quarry, the Main Limestone forms a steeply dipping (70°) massively bedded dolostone, which appears tan, medium to coarse grained with heavily iron-stained colouration and with the occurrence of some accessory minerals like calcite crystals. The state of fossil preservation is extremely poor and no fossil impression were observed in the field.

CEFN-ON AREA: Cefn-On was selected as one of the major sections for two primary reasons, namely, (1) it provides a good reference point for the Main Limestone between Taffs Well and Risca (also previously noted by Dixey and Sibly, 2928, p. 151), and (2) it offers an almost complete development of the Main Limestone Series. The total thickness of the Main Limestone Series was estimated as 920 feet by Dixey and Sibly (1918). According to them, the lower 450 feet comprising Z and part of C_1 zones constituted the dolomite bed, and the next 250 feet embracing C_1 and C_2 zones formed the *"Modiola phase"*, above these, the C_2S (that is, uppermost C_2 and S_1 zones) formed 100 feet of crystalline dolomite and dolomitic limestone, and the remaining portion of the S_1 zone formed a second *"Modiola phase"* of 50 feet. The basal part of the S_2 zone of dolomitized limestone contains 70 feet of oolite which forms the uppermost succession of the Series.

The Lower part of the Main Limestone (ZC_1) dolostones

is partly hidden under the wooded slope of the Cefn-On excarpment. However, their middle part is exposed in the Basic Slag quarry (17408515) near the west side of the Cardiff-Caerphilly railway tunnel, where over 100 feet of the Main Limestone is exposed. These rocks are very dark grey to coarse grained with occasional crinoidal ossicles as the only commonly occuring fossils. The upper part of the Series is also exposed in a small disused quarry (17308530) where abour 20 feet of dolostone characteristically intercalated with fine grained thin bedded dolostone and a relatively thicker, coarse grained crystalline dolostone is found. Finally, the uppermost part of the series is found near the Cefn-On (10008600) in the vicinity of Rudry (Squirrell et al, 1969). Near Draethen, the Main Limestone is exposed again in the Cwm Leyshon quarry (21168686). There, 70 feet of compact, massively bedded dark grey dolostone with associated iron-staining and calcite vein growth dips generally N 40° W.

MACHEN AREA: The basal part of the Series is found in the railway cutting (22138857), where the rocks are chiefly dolomitized limestone with occasional bands of calyey limestone. with argillaceous bands and two zones of dolostone one chert-free above, and other chert-bearing beneath. Somewhat higher horizons of the Series are encountered in the Machen quarry (22208870), where the rocks are thickly bedded, crystalline, medium to coarse grained dolostone characterized by the iron-staining and secondary calcite vein development. Also, some accessory minerals were found from a local fault zone (e.g. galena).

The younger Main LImestone rocks (of S_2 zone) are found in a stream near the Machen quarry, where the rocks are mainly undolomitized oolitic limestone. This locality has geological field significance as it marks the easternmost occurrance of the Seminula zone (S_2 zone) previously shown by Dixey and Sibly (1918).

RISCA AREA: Dixey and Sibly (1918) reported the thickness for the Main Limestone Series as 675 feet made of two parts, the lower 400 feet consist of entirely crystalline dolomite (ZC_1 zone) and the upper 275 feet of predominantly mudstone with intercalcated bands of crystalline dolomite and a thin bed of calcite oolite and calcite-mudstone near the top. The upper part (C_2S_1 zone), according to them constituted the *Modiola phase"*. The dolomitic beds were lacking fssil impressions whereas the calcitic beds contained foraminifera nd ostracods.

George (1956a) noted that the Millstone Grit rapidly oversteps 300 feet of the Series within a distance of 600 yards between the Dan-y-craig and the Risca quarries.

The lower part of the Main Limestone (ZC_1 zone) is observed in the Dan-y-craig quarry (23469090) and the Risca quarry (locally called the "John Adams" quarry [23358980]). In Dan-y-craig quarry, about 300 feet of the Series is exposed, whereas, in Risca quarry slightly over 300 feet thickness is exposed. The basal (60 feet) of the Main Limestone in Dan-y-craig quarry are dark grey, medium grained dolomite with occasional bands of argillaceous limestones or mudstone. These rocks are overlain by 250 feet of coarse grained crystalline dolostone. In the upper part of the quarry, the younger beds of the Series characterized by a fine grained, smooth textured dolostone appear to be the "dolomite-mudstone" (Dixey and Sibley, 1918).

Further north and northeast from Risca, the Main Limestone thins to 200 feet near Ysgbour-newydd. Poor exposures of the series are also found in two small quarries at Blaen-y-cwn Farm, where two feet of dolomite is observed (Squirrell et al, 1969, p. 78(. These workers estimate 40 feet ofMain Limestone exposed at Llanderfel. In Upper Cwmbran, the thickness of Main Limestone is placed between 170-200 feet (Squirrell et al [op. cit.], p. 79), and where the rocks dip at 90°, George (1956a) estimated 150-160 feet of buff-granular dolomite intercalated with quartz pebbles in the upper part. According to George, such conglomerates were found no where in the vicinity (that is, in the south towards Risca and in the north towards Pontypool), and he therefore, attributed their presence to a local shallowing.

PONTYPOOL AREA: The local development of the Main Limestone in the quarries soiuth of Pontypool town was described by Strahan (1909) who estimated the exposed thickness of the Main Limestone as 100 feet. However, some 250-300 feet thickness reported for this area by Squirrell, et al (1969, p. 79).

The Cwm-ynisycoy quarry (29209970) which is chosen as the representative area for Pontypool shows the following 100 feet sequence of the lower pary fo the Main Limestone (ZC_1 zone). The basal 30 feet is made of dark grey, medium grained crystalline dolostone with occasional bands of fine grained black mudstone or argillaceous limestone. The middle bed (30 feet) consists of partially dolomitized oolitic limestone with some irregular argillaceous bands. George (1956a, P. 314) considered this middle to be equivalent to the Gilwern Oolite Group in the Clydach region. The upper (40 feet) bed is predominantly a medium to coarse grained and grey coloured thickly bedded dolomitized limestone.

The Main Limestone is also exposed farther north of Pontypool where it forms prominent escarpment, for example, in the north side of Pontypool Park.

ABERSYCHAN AREA: The lower part of the Main Limestone is found at tow localities, namely, one opposite to the Cwn Afon Inn beyond the former railway station and the other at Abersychan quarry near the Afon Lwydd River. In the former locality, the Main Limestone is 70 feet thick (George, 1956a), whereas, in the latter place the Main Limestone shows 75 feet of massively bedded grey to bluish medium grained dolostone. In view of its dark bluish colouration, the rock is locally known as the "Blue Limestone". Some galena was found in scattered form from a local faulted zone.

VARTEG AREA: About 100 feet of a predominantly crystalline dolostone belonging to the Main Limestone Series is found in a disused quarry locally known as the "Jaw Dan" quarry) located along the railway line near Gallowsgarn. The lower part of the series is made of massively bedded 55 feet of dolostone with nodular-chert (Pl. Ie). This bed is overlain in turn by thin bedded fine grained dolomite (20 feet thick) in which the quartz pebbles are present. The overlying bed (30 feet) comprises massively bedded coarse grained light coloured dolostone. The top part of the series is capped by the Millstone Grit Series.

According to George (1956a, p. 312) the thick bedded coarse grained dolostones are equivalent to the Gilwern Oolite Group of the Clydach region, whereas, the thin bedded dolostone corresponds with the marker beds in the same region. Half a mile north of Varteg Hill, the Craig quarry exposes about 50 feet of the Main Limestone (George, 1956a). In addition, he reported the presence of abundant polyzoans, crinoids and the remains of algal structures in these rocks of the Craig quarry. He correlated with the Oolite Group of the Clydach region. George (op. cit.) pointed out that the Tournasian beds were overstepped by 35 to 45 feet of Millstone Grit Series within a distance of one-half mile between Varteg and Craig.

The Main Limestone is completely absent in the Blorenge area, but recurs near Gilwern Hill.

CLYDACH REGION: The contact between the Lower Limestone Shale (K zone) and the Main Limestone is exposed in the Clydach Valley stream (see also George, 1954) where the bed is about 10 feet. George (1954) subdivided the Main Limestone Series in this region into two major groups, namely, the Lower Oolite Group (Z zone) and the Upper Calcite-mudstone Group (C_2S_1 zone). The thickness of the Lower Group is about 120 feet and it is exposed in the quarries along the Black Rock. The Upper Calcite-mudstone Group is exposed in the Llanelli quarry where it is about 60 feet in thickness. A similar development is also observed in the nearby Cwm quarry. The recent workings in the Black Rock quarry, have exposed fresh, hitherto unseen beds which show part of the Calcite-mudstone. Extimated thickness of these rocks is 100 feet.

The lower part of the Series is made up of dark-gray and fine grained dolostone alternating with light grey oolitic limestone (Pl. If). The upper part of Calcite-mudstone are highly dolomitized limestones which are fine grained, black rocks without any marine fossil impressions. These beds are overlain by the Oolitic Limestone (i.e. the S_2 zone).

Chapter Five

PETROLOGY

CHAPTER FIVE

PETROGRAPHY

Introductory Remarks

As mentioned earlier certain petrographic criteria were used by Dixon (1907) to substantiate his "contemporaneous" origin of dolomite hypothesis. In subsequent years, Sibly (private collection)[8] followed a similar method for the general description of limestone in South Wales. Wood (1941) petrographically identified algal structures or as he designated these as "algal dust" in the Calcite-mudstone of the Clydach region. In general, the petrographic studies relating to the problem of the dolomitization in the Main Limestone Series in the past have been undertaken to establish the time of alteration of these rocks (especially, for example, Dixon, 1907). Bhatt (1972) employed the petrographic method to determine (1) the time of alteration of these rocks under study, (2) the major microfacies indicative of environments, (3) a general pattern of diagenesis, and (4) to obtain supplementary evidence for the chemical data of his study.

The terminology used for the thin section study is based upon the combination of Folk's (1959) classification, Bathurst's (1958, 1959) scheme of the diagenetic fabrics and Friedman's (1965) terminology for the recrystallized rocks.

Mineral Composition

The Main Limestone contains dolomite and calcite as the predominant major minerals, whereas clay, quartz (authigenic and detrital), minor mica, and organic debris make up the subordinate components. The amount of clay is significant in the unaltered limestone and in the argillaceous limestone bands.

CALCITE: Stained calcite crystals appear faint pink to reddish pink with their characteristic obtuse-angled intersection of parting lamellae in a favorably oriented grain. The smaller fraction is present as calcite matrix. In some cases calcite appears as enlarged crystals of crinoids due to "rim cement".

---

[8]This collection is available in the Department of Geology at University College, Cardiff.

In some other cases it appears as "sparry clacite" (Folk, 1959) filling the original organic cavities thus forming "drusy-calcite" mosaic but frequently in the same thin section it also appears as a microcrystalline calcite.

The replacement of calcite by dolomite is observed in the partly dolomitized limestone; on the other hand, the replacement of dolomite by calcite is found in relatively very few cases. Also occasionally, the calcite crystals are seen attacking the quartz fragments.

DOLOMITE: Dolomite appears as colorless idiotopes or hypiotopes of rhombs with a grain size varying between 10 - 200 microns, but 50 to 150 microns is the common range. The dolomite rhombs are characterized by black or brown rims (zonal rims) possibly coated by finely divided iron-oxides. The variation of apparent relief on rotating under the plane polarized light fives them a well-defined pseudo-pleochroism (Braithwaite, 1958).

Moreover, other differences exists between these two carbonate minerals, for example, in relation to calcite, dolomite crystals are usually euhedral or in idiomorphic forms (Milner, 1962; Moorehouse, 1959; Carozzi, 1960; Pettijohn, 1957). The right-angled parting lamellae of dolomite crystals especially in favorably oriented position distinguishes them from calcite and unlike calcite crystals, the dolomite crystals which project into vugs show cruved faces (Moorehouse, 1959). In addition, standard stain and chemical analyses permitted their further distinction.

QUARTZ: Quartz usually appears either as detrital or authigenic secondary minerals, but the former type is more common. Their size varies considerably from 50 - 200 microns and in some cases to 1 — 2 mm or over. The authigenic variety is characterized by their piokolotopic appearance of hexagonal shape.

IRON-BEARING MINERALS: Iron-bearing minerals in the Main Limestones occur as (1) finely divided iron oxides forming the "zonal rim" around the dolomite rhombs, (2) as fine matrix in the intragranular cement and (3) in the secondary calcite veins superimposing the calcite veins (as evident from the royal blue stain). Iron minerals sometimes show the "shrunk-cavities" form of ironsilicates (see also Schmidt, 1965). In general, iron-bearing minerals in the rocks under question commonly occur as disseminated finely divided powder. However, occasionally they do concentrate to a point to form

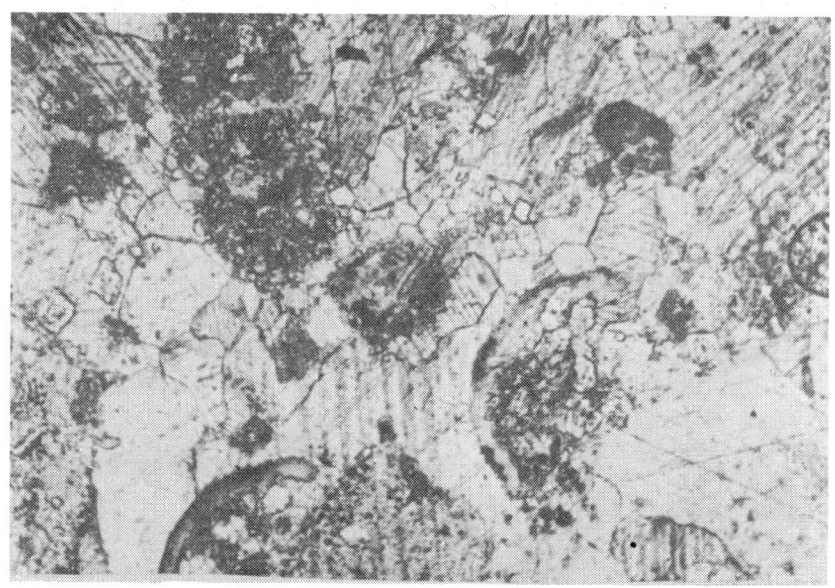

Plate II. Stained calcite crystals showing their characteristic obtuse-angled intersection of parting lamellae in a favorable oriented position (lower right corner). Dolomite rhombs of varying size are seen in the center of the photograph. Detrital quartz crystals are seen in the upper right corner (white grains). Dark color matter represents organic content intricately mixed with clay matter presumably rich in iron and magnesium.

Locality: Tongwynlais near Cardiff, Nicols not crossed, X 16.

ferroan-calcite and ferroan-dolomite varieties, but no where was their concentration sufficient to warrant the formation of ankerite.

CLAY MINERALS: No direct identification of clay minerals was possible under the thin section examination. However, their presence was suspected indirectly by the presence of extremely fine matrix (particularly observed in the argillaceous limestone samples), and in the disseminated flakes of mica but chiefly from the chemical analysis.

ORGANIC MATTER: Organic matter, largely the disintegrated algal structure, appears to be present in the argillaceous limestone and the calcite-mudstone. In general, the original content of such matter may have been destroyed be the dolomitization. Therefore, in the majority of the samples their origin remains obscured.

Cement and Matrix

Both calcite and dolomite cement predominate in the rocks under study. However, occasional presence of their ferroan varieties is encountered in some samples. The cement occurs as sparry calcite and microcrystalline calcite in the undolomitized limestone. The matrix fraction in the relatively unaltered or partly dolomitized limestone remains 20 - 60%. Dominant matrix in such rocks is "microcrystalline calcite" and high matrix percentage of such calcite tends to be common in the lower part of the Series. In highly dolomitized limestone or dolostones, their amount is reduced to 5 - 20%, and it has become relatively more dolomitic in nature. In addition, dolomite matrix and ferroan-calcite or the ferroan-dolomite matrix (identified with the stain technique) were also observed.

Organic Structures and Debris

Crinoid ossicles are one of the most frequent of the few surviving fossil impressions encountered in the dolostones. Crinoids tend to be less common in the occasional beds of the argillaceous limestone (shale-partings) found in the lower part of the Series. However, crinoids do dominate the crinoidal limestones of the basal part of the Series (ZC_1 zone) (Pl. III). In addition to crinoids, brachiopods, bryozoa, and corals occur in the Series. Some foraminifers are observed as one of the organisms resistant to dolomitization (see also Schlanger, 1957).

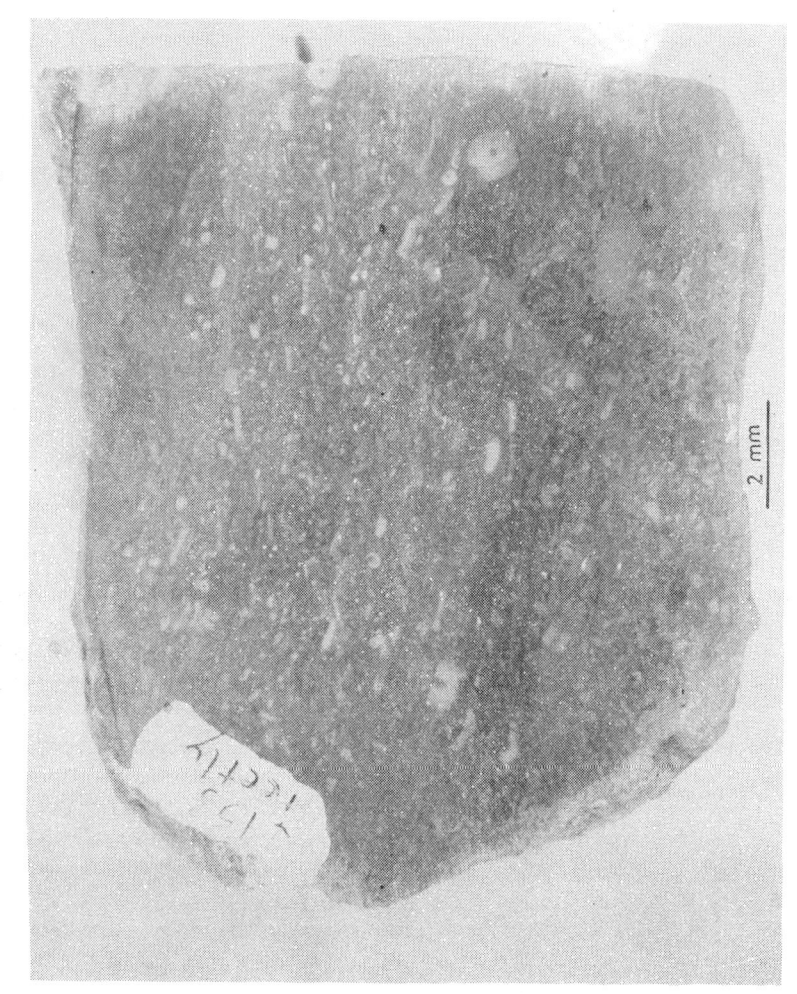

Plate III. Dolomized crinoidal limestone. Locality: Taffs Well Region

Table II. Identification of Algal structures in the dolomitized Main Limestone rocks +

Sample No.	Description	Locality
J512	Possible Siphonales structure. Chlorophyta ((Green Alage) larger than corallina Algae.	Miskin
J513	Possible fragments of corallina.	Miskin
J307	Possible Corallina tips	Tongwynalais
J315	Poorly preseved structure, difficult to identity.	Tongwynalais
J321	Siphanceous type structure	Tongwynalais
J306	A possible axis of an undeterminable object.	Tongwynalais
J130	Possibly Corallina tips (Nemathecia L. S.)	Risca
J722A	Possible Corallina tips	Pwll Du
675	Concentric rings like Lithophyllum or Lithothamnium. Red Algae (Rhodophyta)	Treifl

+ The writer is indebted to Dr. K. Evans-Benson (Department of Botany, University College, Cardiff) who identified these structures.

++ In some additional samples from Machen, Clydach and Taffs Well, Algae like structures were suspected but due to greater degree of obliteration positive identification of such Algae structure was not possible.

Apart from these basal rocks (ZC_1 zone), the general preservation of fossils become extremely poor in the younger beds of the Main Limestone (Upper ZC_1 and C_2S_1 zones) in the Taffs Well region. In the Miskin area, fossils are also relatively better preserved in the lower part of the Series, but they too are the poorly preserved in the younger beds.

Scattered algal remains occur throughout the area of study, but in the middle part of the Series (approximately C_2S_1 zone)[9] they become the most common organic structure occurring in the form of "algal dust" (Wood, 1941). Algae are seen under thin section in the form of broken fragments around the oolites (in the oolitic limestone) and possibly as black organic material at the center of the dolomite rhombs as shown by Taft and Harbaugh (1964). Various forms of red and green algae are found in the rocks under question from the Miskin, the Taffs Well and the Clydach regions. Algal structures in the Clydach region were earlier reported by Wood (1941) and George (1954) who attributed the formation of calcite-mudstone to algal dust of their disintegrated parts. Bhatt (1972) concluded that although algae tends to be abundant in the Clydach region, it now appears that the algae were present throughout the area of the study (Table II).

Sedimentary Structures

Sedimentary structures in the Main Limestone are rather uncommon especially in the highly dolomitized dolostones. Nevertheless, the peel replica and the thin section studies revealed the presence of some structures. They include cross-lamination, parallel lamination, and graded bedding each of which are indicative of a shift in the index of hydraulic energy. The change from parallel-lamination to cross-lamination was observed in the lower part of the Main Limestone Series. George (1954) has reported the occurrence of sun-cracks and seaweed frond features in the calcite-mudstone from the Clydach region, which he attributes to shallow water conditions of less than three feet deep.

Major Microfacies

Six major microfacies in the Main Limestone rocks were

---

[9]Calcite-mudstone group in the Clydach region (George, 1954).

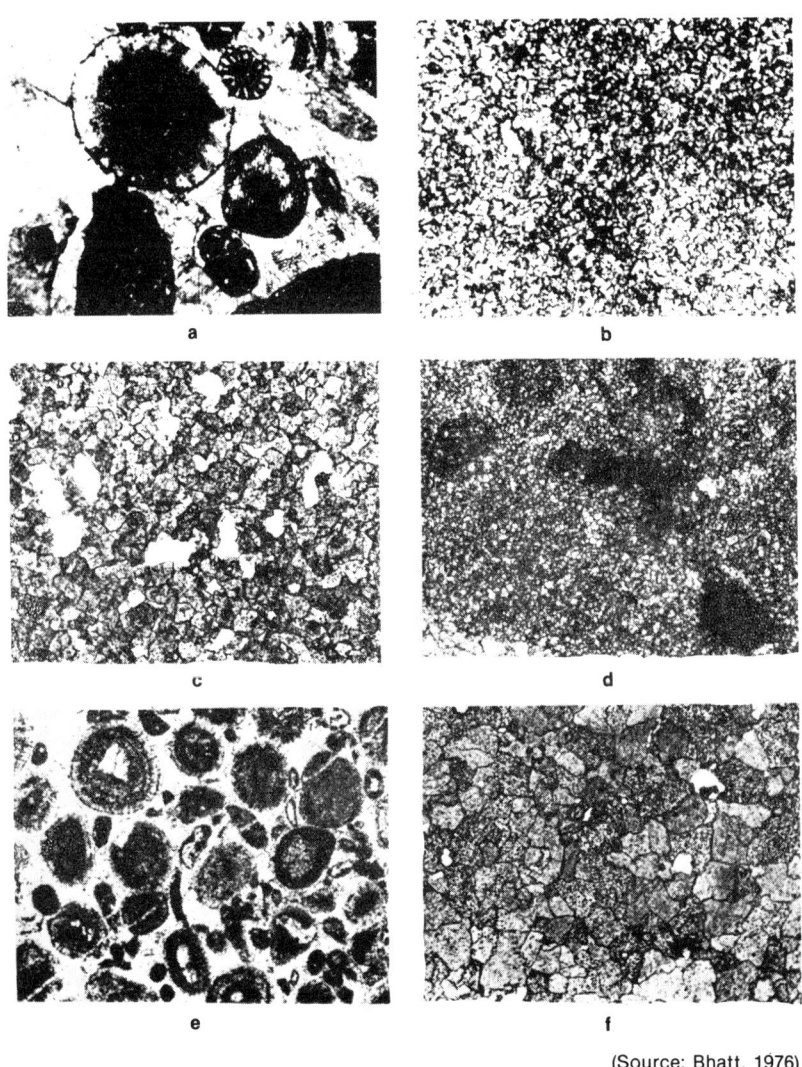

(Source: Bhatt, 1976)

Plate IV. Major microfacies as observed in the Main Limestones of South Wales. See the text for further explanation (Section 5-5).

revealed under thin section examination and are defined as follows: (Bhatt, 1972, 1976).

Microfacies I BIOSPARITE: Refers to the undolomitized fossiliferous limestone in which the allochems of crinoids, bryozoa corals, and barchiopods are cemented by sparry calcite cement.

Microfacies II DOLOBIOMICRITE: Refers to the original organically formed material (calcareous muds) which metasomatically altered into dolostone in a relatively stable condition.

Microfacies III QUARTZ-BEARING DOLOBIOMICRITE: Refers to the originally organically formed material (calcareous muds) which metasomatically altered into dolostone and such process took place in relatively unstable conditions as evident by the enclosed detrital quartz fragments.

Microfacies IV DOLOMICRITE: Refers to the original fine-grained (organic or inorganic) $CaCO_3$ material which altered syngenetically or diagenetically into dolostone without being subject to significant grain growth process.

Microfacies V OOSPARITE: Refers to the unaltered and partly altered limestone in which the chief component ooliths (and pisolites) are cemented by sparry calcite.

Microfacies VI DOLORUDITE: Refers to the presence of very coarse grained dark color dolomite rhombs surrounded by light color fine matrix. Their varying grain size indicated either various stages of "grain-growth" or in part they may be the products of "evaporative-solution" or both.

BIOSPARITE (microfacies I): The rock is predominantly made of allochems of crinoids, bryozoa, brachiopods, corals, and foraminifera all cemented by sparry calcite. However, some portion of the rock is occupied by the microcrystalline cement. The size of the clasts vary from 100 microns to 1 to 2 cm with 0.5 to 1.0 mm common. Poikiolotopes of dolomite euhedra of less than 100 microns are present in scattered fashion in an amount not exceeding 5% (vol.%) of the total area of the slide surface in some samples. These rhombs are confined to the fine microcrystalline cement (Pl. IV a).

DOLOBIOMICRITE (microfacies II): The rock is characterized by the presence of hypiotopes (occasionally idiotopes or xenotopes) of dolomite with a size of 50-150 microns. There is an almost complete absence of quartz fragments in the majority of cases. The preservation of fossil impressions is poor and usually shown by the "drusy calcite" infilling of the former organic cavities (Pl. IV b).

QUARTZ-BEARING DOLOBIOMICRITE (microfacies III): The texture of the rock is characterized by the presence of hypiotopes to xenotopes of dolomite varying in size from 50-200 microns. Usually such rocks are associated with detrital or to a lesser extent with authigenic quartz fragments of 1% to 5%. In some cases, the quartz is observed to be replaced by the carbonate minerals as shown by the "quartz-frosting" texture. Fossil impressions are very poor in this M. F. III.

The microfacies III is distinguished from the microfacies II in three major respects, namely: (1) higher percent of quartz fragments, (2) varying grain size, and (3) near absence of fossil impressions including the drusy-calcite cavity fillings of a former organic chamber of a cavity.

DOLOMICRITE (microfacies IV): The rock is characterized by the presence of very fine grained microcrystalline dolomite cement of less than 100 microns size with 30-50 microns being the common size. The identification of such fine material under thin section with a low power was difficult but the higher magnification revealed such material to be dolomite matrix. There is a complete absence of fossil impressions. The dolomite rhombs appear as idiotopes of perfect shape but small size, usually in clustered form.

OOSPARITE (microfacies V): The rock is predominantly made of oolite grains cemented by sparry calcite (Pl. IV e). The size of the ooides vary between 0.5 to 2.0 mm but also sizes 2.0 mm (pisolites) are common. The ooides are rounded, subrounded or cylindrical in shape with concentric structures at their rims which are characterized by the encrusting algal structures or preferably the "Algal circumcrusted rims" (Wood, 1941; and Taft and Harabaugh, 1964).

DOLORUDITE (microfacies VI): The rock is characterized by the presence of xenotopes of dolomite in varying sizes of 0.5 mm to over 2.0 mm and even higher (Pl IV f). In appearance these are the "pseudobreccis" of Bathurst (1959) characterized by the grain-growth mosaic. In general, the fragments appear light and the adjoining fine grained matrix dark. Quartz fragments are also found in such microfacies.

Fig. 5.

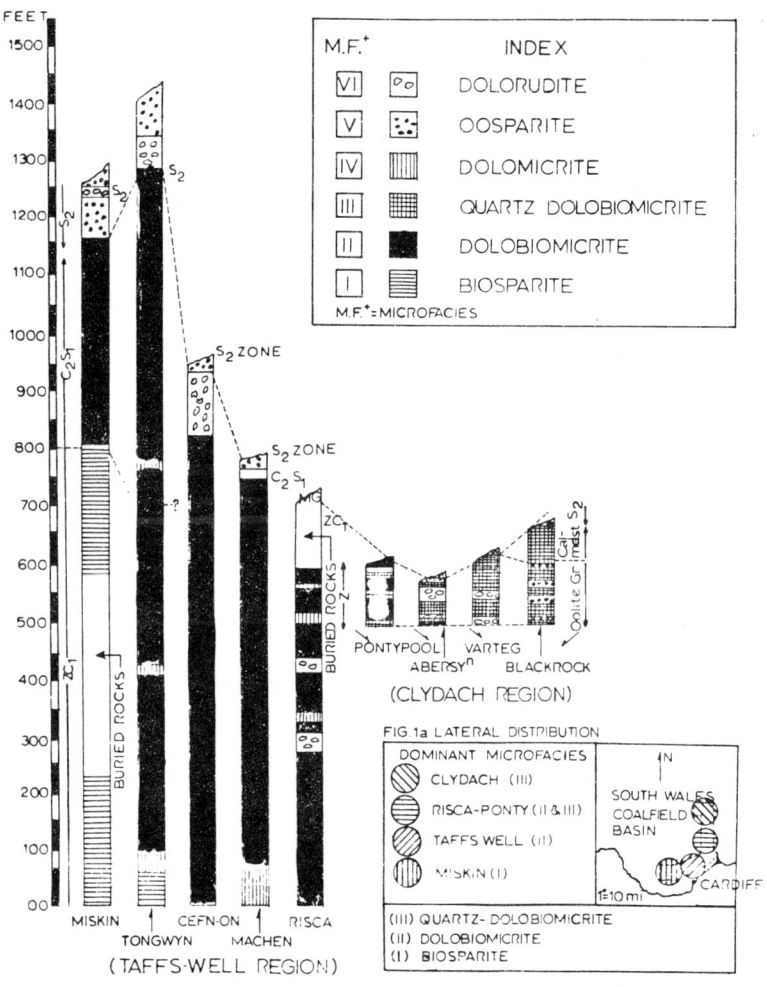

Distribution of Microfacies

In terms of their vertical distribution, the biosparite dominate in the basal bed (approximately Lower ZC_1), the dolobiomicrite (M.F. II) and the quartz-bearing dolobiomicrite (M.F. III) in the middle beds (approximately Upper ZC_1 and C_2S_1 zones) and the oosparite (M.F. III) in the middle beds (approximately Upper ZC_1 and C_2S_1 zones) and the oosparite (M.F.V.) in the upper bed (i.e. in the S_2 zone) of the Main Limestone Series.

In terms of their lateral distribution, the biosparite (M.F. I) tends to dominate in the general area of Miskin, the dolobiomicrite (M.F. II) in the Taffs Well region and the quartz-bearing dolobiomicrite (M.F. III) in the Clydach region. Interestingly, on the other hand, the area between Risca-Pontypool shows overlapping of the microfacies II and III, thus suggesting a transitional position between the Taffs Well and the Clydach regions (Fig. 5).

Chapter Six
GEOCHEMISTRY

Table III. Ca/Mg ratios in the Main Limestone Series of South Wales.

Taffs Well region

Miskin		Tongwynlais (Bhatt 1973)				Cefn-On	
Sample No.	Ca/Mg	Sample No.	Ca/Mg	Sample No.	Ca/Mg	Sample No.	Ca/Mg
J501 (top)	115	J327 (top)	30	J313	1.9	J200 (top)	2.8
J502	10.5	J326	2.3	J312	1.8	J201	2.2
J503	1.30	J325	110	J311	1.7	J202	1.8
J504	3.80	J324	56.2	J310	1.7	J203	2.7
J505	2.40	J323	1.6	J309	5.9	J204	2.0
J506	2.10	J322	1.6	J308	4.2	J204A	1.6
J507	3.90	J321	1.6	J307	9.5	J205	1.8
J508	2.40	J320	1.7	J306	1.0	J206	2.0
J509	29.0	J319	1.6	J305	1.0	J207	1.7
J510	45.0	J318	1.4	J304	11.0	J208	1.6
J511	20.0	J317	1.6	J303	20.0	J208 (bottom)	1.6
J512	1.20	J316	1.9	J302	14.0		
J513	10.9	J315	1.5	J301	14.0		
		J314	1.6				

Risca-Pontypool region / Clydach region

Risca		Pontypool		Blackrock		Other areas	
Sample No.	Ca/Mg	Sample No.	Ca/Mg	Sample No.	Ca/Mg	Sample No.	Ca/Mg
J612 (top)	1.70	J133	2.1	J822 (top)	235		
J611	1.76	J132	40.0	J820	5.6		
J610	2.50	J131	2.9	J819	42	*Abersychan*	
J609	1.69	J642	2.2	J818	1.6	J633	2.8
J608	1.76	J641	2.1	J817	3.4	J632	1.8
J607	1.85			J816	2.6	J631	2.0
J606	3.69			J815	4.3	J630	3.3
J605	2.60			J814	1.6		
J604	2.07			J813B	3.9	*Varteg*	
J603	2.07			J813A	1.7	J710	6.2
J602	2.24			J812	1.7	J708	5.8
J601	5.00			JS11	1.5	J830B	47.0
				J805	99		
				J804	3.8	*Pwll Du*	
				J803	3.80	J722A	12.8
				J802	—	J722B	5.0
				J801	2.6		
				J800	400		

(Source: Bhatt, 1973, 1976)

CHAPTER SIX

GEOCHEMISTRY

Introductory Remarks

The geochemical study of the Main Limestones include quantitative determination of common ratios: Ca and Mg, K and Na, Fe and Mn, and Ti and Al.

Ca/Mg Ratio Distribution

Daly (1907, 1909) pioneered the use of Ca/Mg ratio method by demonstrating with time a progressive increase of Ca/Mg ratio of the sediments collected from the United States, Canada, and Belgium. His work was later supported by Vinogradov. Ronov, and Ratynsky (1952 in Graf, 1960) and by Chilingar (1956b). The mean value of the ratio was only 2 in Pre-Cambrian, in the mid-Paleozic it was under 10, but increased to 90 in Cretaceous, before finally decreasing to the present day 40 (Fairbridge, 1964). The Cretaceous anamoly is attributed to the evolution of low Mg-calcite carbonate organisms, especially the foraminifera.

In terms of vertical distribution in the Main Limestone Series, the Ca/Mg ratio indicates relatively *variable* behavior in the lower part (approximately ZC_1 zone), *uniform* behavior in the middle part (approximately C_2S_1 zone) and a return to *moderately variable* behavior in the upper part (basal S_2 zone). For example, in the Tongwynlais section, the basal beds of the Main Limestone Series show wider range of the Ca/Mg ratio from 1:1 to 50:1 and very high 100:1, whereas in the middle part it (Ca/Mg ratio) remains within a narrow range of 2:1, but later in the upper part of the Series (basal S_2 zone) it tends to go up from 5:1 to 50:1 and even higher (except for local vein-dolomite zones where it falls considerably low in relation to the unaffected adjoining beds). (See also Bhatt, 1973).

The lateral distribution of Ca/Mg ratio suggests a regional pattern (Fig. 5). For example, the Ca/Mg ratio behavior in Miskin and more particularly in the Taffs Well region shows relatively less variable that is a moderately uniform behavior, in the Clydach region a relatively greater variability and the

Table IV. Magnesium distribution in major sections from the Main Limestone rocks of South Wales.

Miskin*		Tongwynlais				Cefn On		Risca		Varteg*	
Sample No.	Mg%	Sample No.	Mg%	Sample No.	Mg%	Sample No.	Mg%	Sample No.	Mg%	Sample No.	Mg%
J501	1.4	J327	5.2	J313	45.1	J200	43.1	J612	49.1	J830B	3.4
J502	14.1	J326	41.4	J312	47.0	J201	42.2	J611	48.3	J708	22.2
J503	tr.	J325	1.5	J311	48.0	J202	46.2	J610	39.7	J710	34.7
J504	30.2	J324	2.8	J310	43.6	J203	43.1	J609	49.7		
J505	41.1	J323	50.9	J309	27.3	J204	45.1	J608	48.1		
J506	44.2	J322	50.3	J308	33.6	J204	50.0	J607	47.1		
J507	29.5	J321	50.0	J307	14.8	J205	48.2	J606	30.8		
J508	40.3	J320	49.4	J306	62.2	J206	45.1	J605	38.8		
J509	5.3	J319	50.0	J305	62.2	J207	49.5	J604	44.2		
J510	3.5	J318	53.1	J304	1.5	J208	50.7	J603	44.2		
J511	7.6	J317	50.1	J303	0.82	J209		J209 5017		J602	42.4
J512	57.8	J316	47.4	J302	1.2			J601	24.8		
J513	13.1	J315	51.3	J301	10.5						
		J314	50.9								

* Note the relatively greater variation of Mg content in the Miskin and Varteg regions.

(Source Bhatt, 1976)

Figure 6. Bimodal distribution of Mg content in the Main Limestone (after Bhatt, 1976).

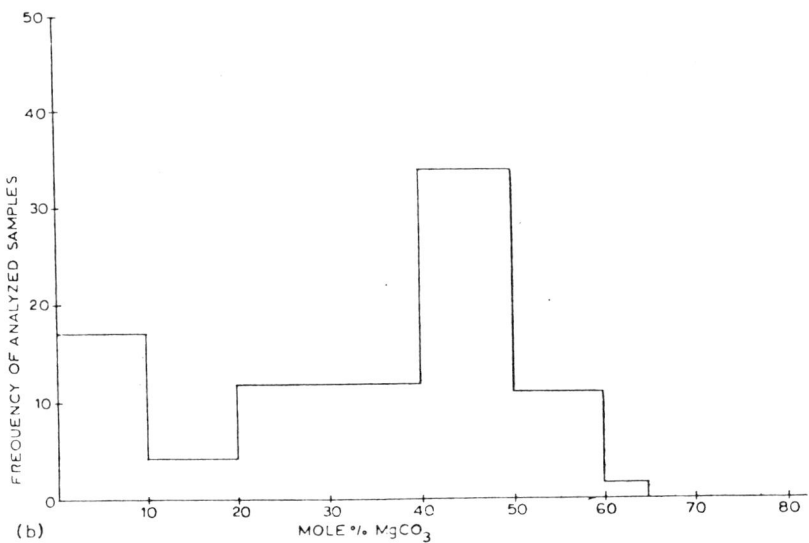

Risca-Pontypool region in general is characterized by moderately-variable behavior.

Magnesium Distribution

On plotting the frequency of samples analyzed against the mole per cent of $MgCO_3$ the general distribution of dolomitization intensity in the Main Limestone Series yields a bimodal pattern of dolomitization (Fig. 6).

By using the petrographic method as a complimentary to the chemical method, Bhatt (1974) observed that the majority of the samples having a Mg-content less than 2.5 mole % $MgCO_3$ showed no sign of dolomitization, therefore he considered these samples as *unaltered limestone*. Those samples showing magnesium content between 2.5 to 10.0 mole % $MgCO_3$ were regarded as *moderately dolomitized limestone* and those between 10 to 40 mole % $MgCO_3$ were taken as *highly dolomitized limestone*, only those samples with 40-50 mole % $MgCO_3$ were categorized as the *true dolostone* of this study. A few samples having value of above 50 mole % $MgCO_3$ were classified as *magnesian-dolostone*. The majority of the samples so analyzed showed values between 40-50 mole % $MgCO_3$ or the "true dolostone" in the Main Limestone rocks, thus indicating a remarkably advanced state of dolomitization attained by the Series under study.

Bimodal histograms of magnesium content in the carbonate rocks from the geologic column have been shown previously by some workers elsewhere; for example, in the United States (Chave), 1954b, p. 548), in Russia (Ronov *et al*, 1956; Nikoloy *in* Graf, 1960), and in Germany (Schmidt, 1965). The bimodal distribution of the magnesium content can be attributed to: (1) alternate deposition of lime facies and dolomite facies, (2) selective dolomitization of fine material from the partially dolomitized lime facies (Schmidt, 1965), (3) statistical reinforcement on account of the presence of two extreme mixtures of the carbonates, namely, the unaltered limestone and the relatively highly dolomitized dolostone, and (4) periodic-climatic variations determining biochemical activities, for example the calcareous Algae with their high or low participation cause the "waxing" (Fairbridge, 1967) or the "wanning" (Sarin, 1962) periods.

It is believed that the above mentioned reasons may well be applicable to the Main Limestone Series in which the

Table V. Geochemcial data on the Main Limestone rocks in Tongwynlais section, Cardiff district.

Sample No.	Ca/Mg*	Mole percent MgCO$_3$	Mn/Fe	K/Na	Clay%
J327 (Top)	30.0	5.2	tr.	0.30	tr.
J326	2.3	41.4	0.43	0.13	5.0
J325	110.0	1.5	tr.	0.12	tr.
J324	56.2	2.8	tr.	0.12	29.0
J323	1.6	50.9	tr.	0.15	14.0
J322	1.6	50.3	0.33	0.66	8.2
J321	1.6	50.0	0.32	0.22	4.5
J320	1.7	49.4	0.45	0.25	14.2
J319	1.6	50.0	0.20	0.25	7.6
J318	1.4	53.1	0.016	0.28	18.9
J317	1.6	50.0	0.40	0.80	6.0
J316	1.9	47.4	tr.	0.66	13.6
J315	1.5	51.3	0.62	0.33	16.5
J314	1.6	50.9	tr.	0.22	7.3
J313	1.9	45.1	0.40	0.50	27.5
J312	1.8	47.0	tr.	0.66	15.9
J311	1.7	48.0	tr.	0.40	tr.
J310	1.7	43.6	tr.	0.22	30.0
J309	5.9	27.3	0.21	0.40	14.7
J308	4.2	33.6	0.01	0.97	62.7
J307	9.5	11.8	0.28	0.28	18.0
J306	1.0	62.2	0.54	0.76	30.4
J305	1.0	62.2	0.24	1.0	no data
J304	11.0	1.5	0.01	0.72	47.8
J303	20.0	0.82	0.20	0.22	no data
J302	14.0	1.2	0.17	0.56	32.4
J301	14.0	10.5	0.17	0.72	36.9

* Data obtained from Bhatt (1973).

bimodal distribution of Mg-content is also present (Fig. 6).

K/Na Ratio Distribution

Some investigations of potassium and sodium ratio in marine sediments have been recorded by Welby (1958), Heir and Adams (1964). The explanation for the relative low amount of potassium in relation to sodium is observed by other workers (Grim, 1958; Nicholls and Loring, 1960; Rankama and Sahama, 1950). Welby (1958) suggested a correlation between lithology and K/Na ratio in the sediments of the Gulf of Mexico; according to him, sediments with larger amounts of calcium corresponded with low K/Na ratio. Welby (op. cit.) attributed such relation to the fact that higher sodium in the calcareous sediments possibly resulted from biogenic action as certain organisms had the capacity to concentrate sodium. However, the problem appears to be more complex than this since the fate of both potassium and sodium, other than biogenic, depends upon a host of factors ranging from megascopic scale, for example, proverance to sub-microscopic; for example, the nature of ionic radii.

The K/Na ratio in the Main Limestone Series from a majority of samples analyzed indicate a fairly low value of less than 1. The regional distribution of such ratio in the study area indicates an interesting pattern, for example, K/Na ratio tends to be moderately uniform in the Miskin and the Taffs Well regions whilst remaining greatly *variable* in the Clydach region. The Risca-Pontypool region represents an intermediate pattern as their ratio behavior is *moderately variable*. The general range of K/Na ratio in Miskin is between 0.11 and 0.25; in the Taffs Well region it is between 0.20 and 0.80; and in the Clydach region it varies widely between 0.04 and 2.3 with remarkable alterations. In the case of Risca, the lower beds indicate the ratio values to be over 1, whereas in the upper beds it falls below 1, thus suggesting an intermediate behavior in their pattern compared to the Clydach and the Taffs Well regions.

In the Main Limestone Series, the K/Na ratio corresponds inversely with the Ca/Mg ratio in some samples analyzed (Fig. 7). This relationship may be explained:
    (1) Possibly due to the biogenic activity as pointed out by Welby (1958) mentioned in the earlier discussion.
    (2) the post-depositional diagenetic changes brought into the sediments (Nicholls and Loring, 1960),
    (3) the base-exchange properties of clay mineral (Von

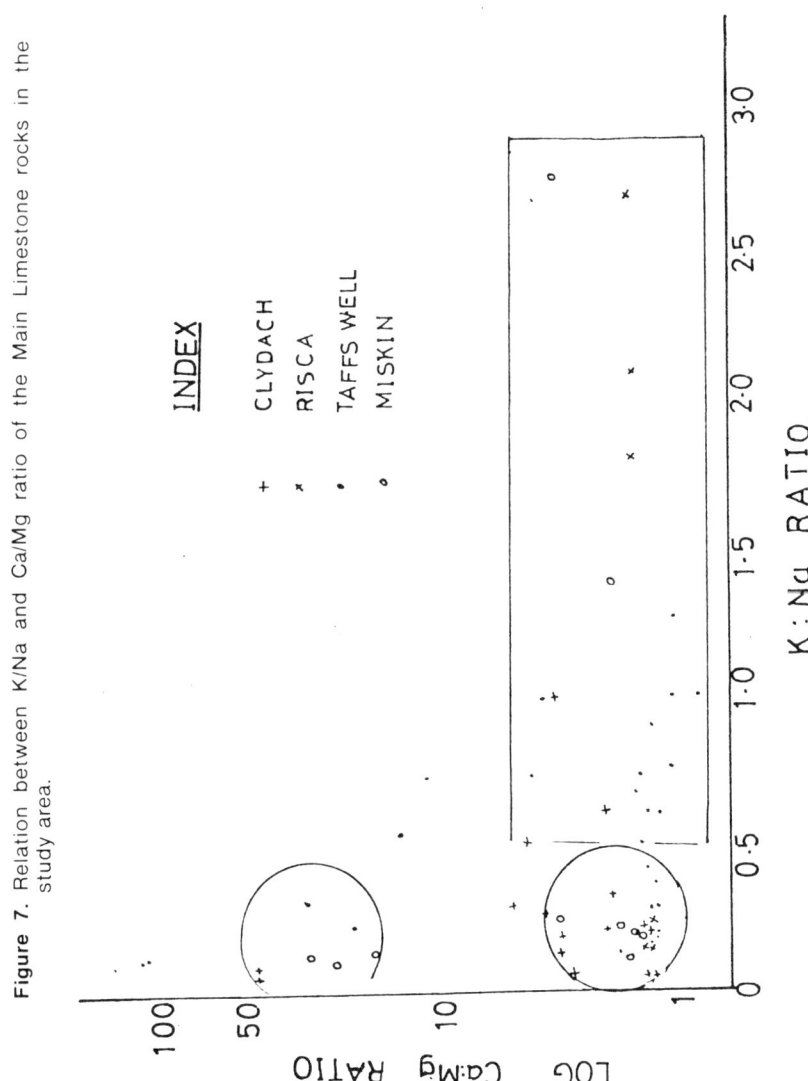

Figure 7. Relation between K/Na and Ca/Mg ratio of the Main Limestone rocks in the study area.

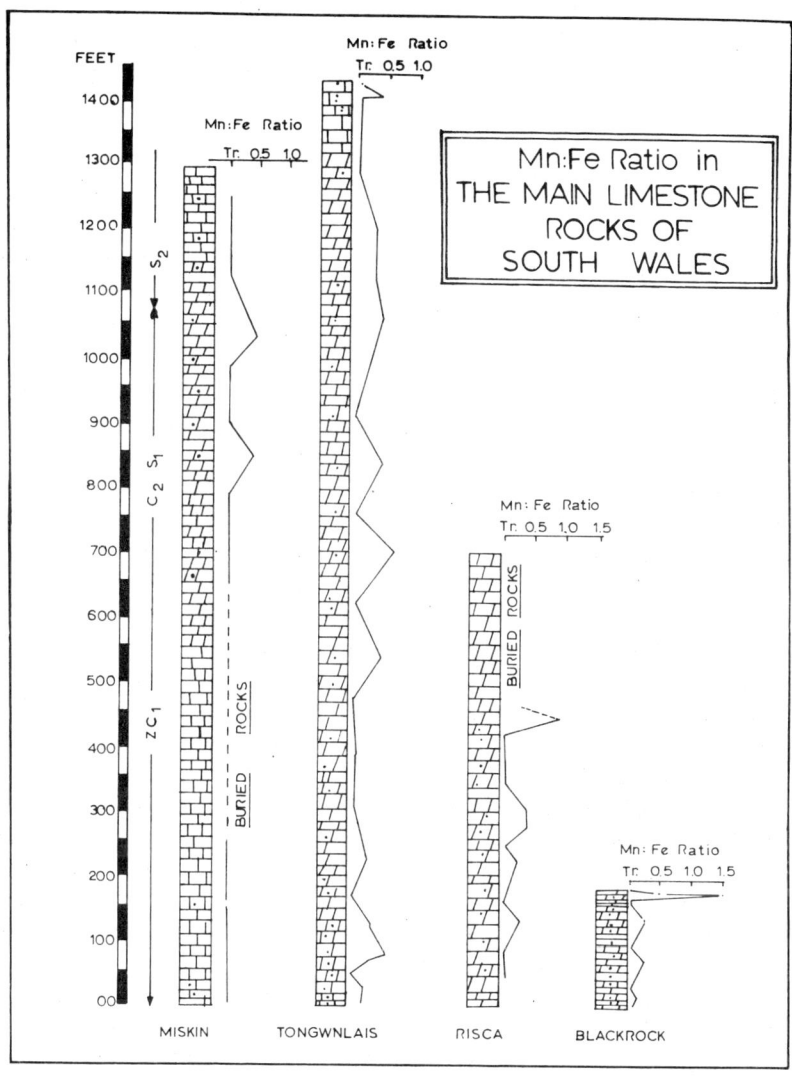

Figure 8. Mn/Fe ratio behavior in major stratigraphic sections in S. Wales.

Engelhart, 1961) and quite possibly due to
(4) the type of dominant clay mineral present in the sediment, for example, either as Na-rich kaolinite or Mg-rich montmorillonite.

Mn/Fe Ratio Distribution

Iron and manganese irons respond slightly differently to changes in pH and Eh. For example, the calculations and experimental results of Krauskopf (1957) and Garrels et al (1960) demonstrated that at any given pH, iron oxides and hydroxides precipitate at lower redox (Eh) than do the manganese oxides. Similarly, under fixed Eh, iron precipitates as oxides at a considerably lower pH than manganese. However, under the environment, where the canbonates are depositing, both of these elements will precipitate simultaneously since their rates of solubilities are not so different. That is, both of these elements will be separated under most exogenic environments in view of the lower solubilities of the iron oxygen compounds (cf. Degens, 1965).

In general, the Mn/Fe ratio indicates values which fall below 1:1 in the majority of the samples analyzed from the Series under study (Fig. 8). For example, the Mn/Fe ratio in the Taffs Well region is between trace to 0.5:1 with moderately uniform vertical distribution; in Miskin it is negligible, whereas in the Clydach region, although the general value of the ratio remains between trace and 0.5, there is a considerable fluctuation (vertically speaking, see Fig. 8). Similarly, Risca indicates the pattern close to the Clydach region but with less vertical variation, hence marking an intermediate behavior between the Taffs Well and Clydach regions.

The Mn/Fe ratio shows a regional trend roughly corresponding to the K/Na ratio (Bhatt, 1972). Also using Friedman's (1969) work as the basis of interpretation, the plot of Mn (ppm) against Fe (ppm) does indeed suggest a wide sea becoming restricted with time (that is, from ZC_1 to C_2S_1 zone; Fig. 9) as evident from the low values of Mn/Fe ratio in the lower beds passing into relatively higher values in the middle part.

According to Rankama and Sahama (1950, p. 647) the Mn/Fe ratio in sea water, for example, is 5:1 for the carbonate-bearing solution, and 1:1 for the sulfate-bearing solution. The Mn/Fe ratio from this study indicates values closer to the latter kind of solution as the ratio is below 1 in the samples

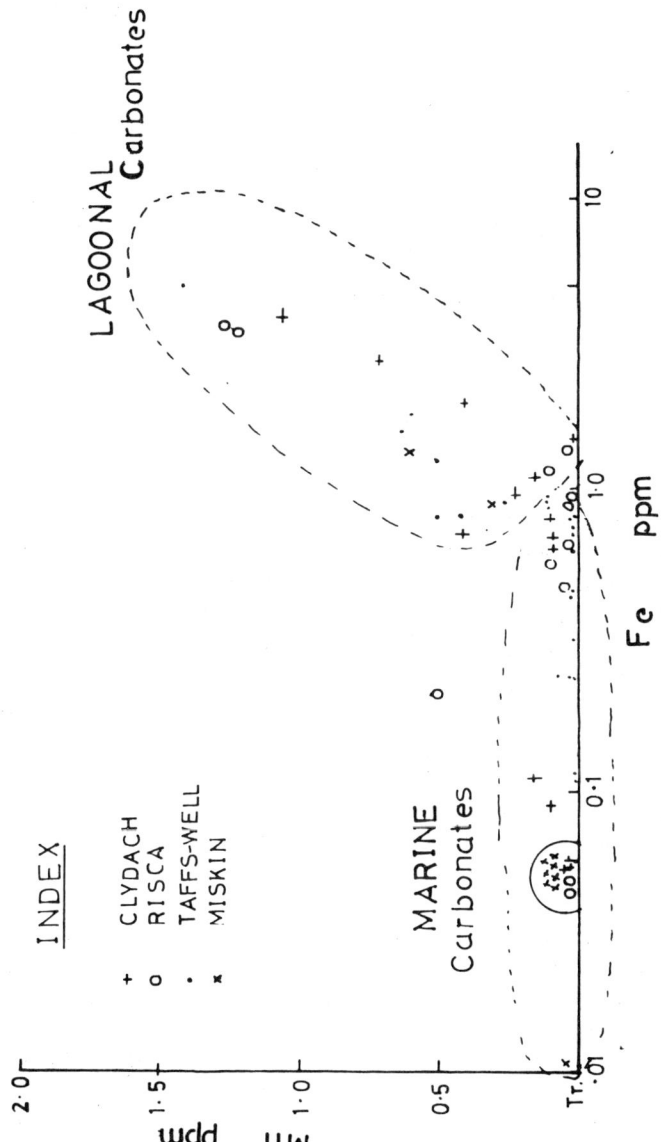

Figure 9. Trace element distribution showing the environment of deposition of the Main Limestone rocks in the study area (Based after Friedman, 1969).

analyzed. It can be argued that the lower-than-1 ratio value may well have been greatly due to the activities of the sulfate-reducing bacteria, e.g. *Desulfurvibrio-desulfuricans* which could have preferably fournished into the then existing sulfate-reducing solution of the Main Limestone sediments.

In summary, it appears that the results of the Mn/Fe ratio of this study, if interpreted in terms of existing knowledge (Ranakama and Sahama, 1950) shows that all information so obtained is consistently point towards the presence of sulfate-bearing sea water medium which was progressively becoming higher in sulfate content with time as the wide and deep Sea became more shallow and restricted. This interpretation was confirmed by the discovery of evaporite minerals in the Main Limestone Series (Bhatt, 1975).

Ti:Al Ratio Distribution

Investigations of the titanium and aluminium ratio has been recorded by a few workers (Correns, 1954, Goldberg and Arrehenius, 1958; Migidisov, 1960 *in* Chilingar *et al*, 1967). According to Goldsberg and Arrehenius (cf. Chester, 1965, p. 33) there was no increase in Ti/Al ratio in sediments containing varying amounts of biogenous components, implying that the biosphere was not the determinable factor for titanium. Migidisov (1960 *in* Chilingar *et al*, 1967) investigated the behavior of Ti/Al ratio in the sediments from the Russian Platform and interpreted the general increase in the ratio as an indication of the oncoming of a humid climate.

Ti/Al ratio study of the Main Limestones investigated by Bhatt (1974) revealed fairly low values (Table VI). The three selected samples taken at a 25 foot interval in a 75 foot vertical section at Steetly quarry revealed a gradual upward increase in TiO_2/Al_2O_3 ratio.

Although the change in their ratio is not of any great magnitude, it does indeed provide a gradual linear change (Table VII). A parallel increase of $SiO_2\%$ in the same samples was also observed.

Bhatt (1974) used Ti/Al ratio to interpret the existence of humid climatic conditions which were slowly developing near the end of the Main Limestone sedimentation. This is also supported by the lack of extensive evaporite deposits in the area, (Bhatt 1975) possibly due to such humid conditions which precluded their preservation. For example, Shinn *et al*

Table VI. $TiO_2:Al_2O_3$ Ratio Distribution in Some Selected Samples of the Main Limestone Rocks, Cardiff District.

Sample No.	TiO_2/Al_2O_3 ratio	Ca/Mg ratio
J118	0.069	--
J129	0.068	--
J143	0.052	--
J148	0.079	3.7
J150	0.098	3.9
J151	0.100	4.4
J160	0.064	High

J148, J150 and J151 were taken at 25 ft. interval in order of their decreasing age. Also notice the correlation between Ca/Mg and Ti/Al ratios in these samples.

(Source: Bhatt, 1974)

Table VII. Trace elements distribution in Main Limestone rocks.

Sample No.	$SiO_2\%$	$TiO_2\%$	$Al_2O_3\%$	$FeO\%$	$Mn_2O_3\%$	$Cr_2O_3\%$	$BaO\%$	Locality
JB 118	1.00	0.03	0.43	0.95	0.12	0.02	0.19	Taffs Well
JB 129	1.67	0.05	0.72	1.02	0.12	0.02	0.005	
JB 143	6.09	0.09	1.72	1.74	0.16	0.01	0.85	Taffs Well
JB 148	2.03	0.07	0.88	1.20	0.10	0.03	0.005	Steetly
JB 150	1.83	0.05	0.51	1.59	0.20	0.02	0.06	Steetly
JB 151	1.41	0.05	0.48	1.61	0.22	0.02	0.04	Steetly
JB 160 (oolitic)	5.07	0.07	1.08	5.69	0.45	0.02	0.005	Gilwern hill (Clydach)

Remarks: JB118 is argillaceous limestone
JB129 and JB160 are clay-free limestones
J143 is dolostone proper (see pp.66-67)
JB148-150-151 are dolomitized limestones

(1965) showed that the humid condition in the Recent sediments of the Bahamas does not permit the preservation of evaporite deposits.

BaO% Distribution

Geochemical studies show that (1) the amount of barium in dolostone is normally one-fifth of that in limestone (cf. Wolf, et al, 1967) and (2) the aragonite lattice structure favors the inclusion of cations like Sr, Pb, Ba, and low Mn. Moreover, this relationship is reinforced by the high pH and temperature at which aragonite precipitates. On the other hand, calcite precipitates at relatively low pH and low temperatures and tends to have low Ba, Sr, Pb, and high Mn (Goldberg and Arrehenius, 1958). Thus the barium per cent in carbonates may represent indirectly the presence of the original carbonate mineralogy, more particularly between calcite or aragonite.

BaO% in some selected samples from the Main Limestone Series revealed interesting results (Table VII). In most commonly occurring rock-type "dolostone" (as represented by the sample JB143) the unaltered oolitic limestone (i.e. not subject to dolomitization shows BaO as 0.005. It seems the higher value of BaO in dolostone may well reflect the nature of their original mineralogy, which was possibly characterized by aragonite. This contention is also supported by the low Mn content of 0.16% (JB143) in relation to MnO% of 0.45 (e.g. JB160 in Table VII).

Interestingly, the amount of BaO% in the clay-rich limestone samples (e.g. JB118 with 0.19% in comparison to the clay-free limestones is higher (0.005% in JB160 and JB129). It is believed that the high value of Ba in the clay-rich sample (JB118) suggests its absorbtion by the clays[11] as suggested by Ranakama and Sahama (1950), but the consistent low BaO% in the clay-free (unaltered) limestones (e.g. JB160 and JB129) and the partly dolomitized limestones (e.g. JB148-150-151) perhaps indicate calcite as their original mineral.

---

[11] In the Main Limestone Series, the clay-rich limestone beds are far less common than those of the dolostones. Therefore, the absorbtion of Ba by the clays can be regarded as a less important factor.

Chapter Seven

HISTORY OF DEPOSITION

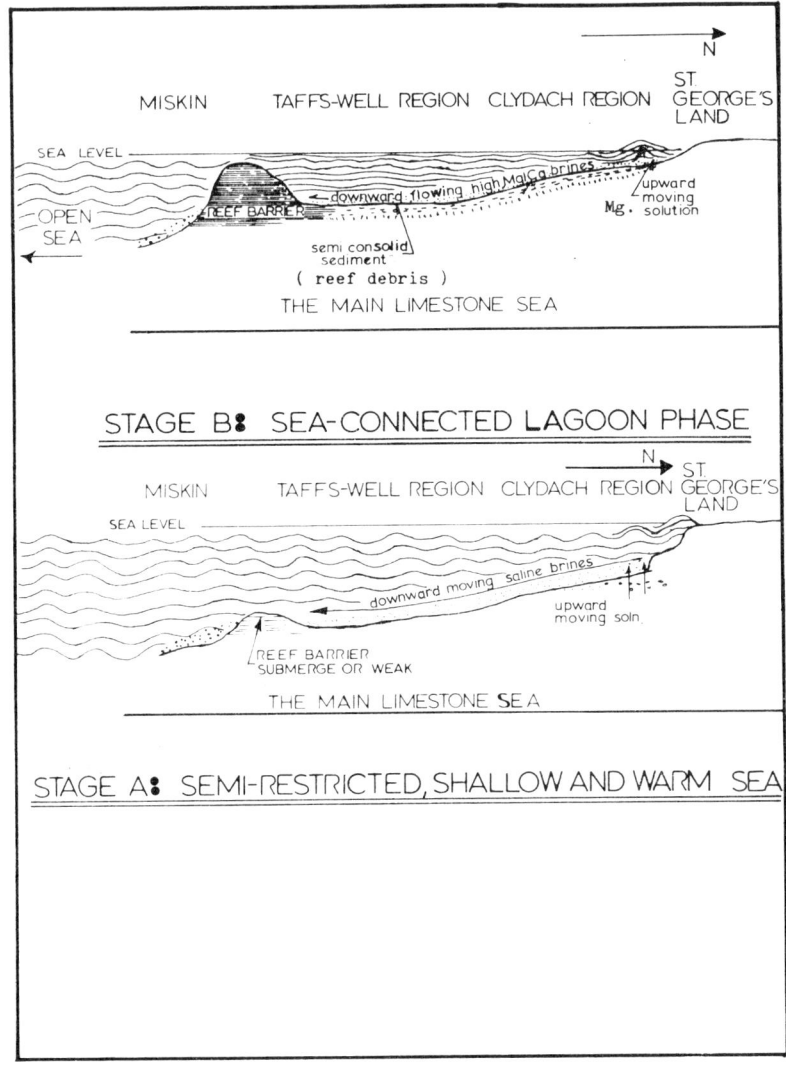

Figure 10. Origin of Dolomization of the Main Limestone rocks in South Wales.

CHAPTER SEVEN

HISTORY OF DEPOSITION: SEQUENCE OF EVENTS

Introductory Remarks

Considering the complex nature of the paleogoegraphy which existed during Carboniferous time expecially in the Southwest Province (George, 1958), the history of deposition of the Main Limestone rocks can only be outlined in terms of simplified stages of development. These stages are: (1) *Stage A* of partial submergence, (2) *Stage B* of partial emergence, and (3) *Stage C* of partial resubmergence. For the sake of brevity the Stages are roughly equated with stratigraphical zones thus: *Stage A* with the Lower ZC_1, *Stage B* with ZC_1-C_2S_1, and *Stage C* with the basal part of the S_2 zones (Table I).

Stage A: Partial Submergence Phase[12]

The history of deposition of the Main Limestone in the earliest part of the Series (Lower ZC_1 zone) was characterized by two closely interrelated controlling features, namely: (1) the presence of a relatively deeper and wider sea, and (2) the prevailing moderate instability of the site of deposition resulting in an unbalanced relation of subsidence and sedimentation. The former feature of the deep and wide sea conditions permitted marine organisms like crinoids, bryozoa, corals, brachiopods, foraminifera to flourish more abundantly in the Miskin and the Taffs Well region but they were becoming less abundant towards the Clydach region where presumably the environmental conditions (*e.g.* salinity) were unfavorble for their survival.

The general instability caused frequent alterations of the carbonate rock deposition, where the limestone and dolostone facies signified temporary transgressive and regressive phases respectively. The alternate deposition of the crinoidal limestone with the clay-rich limestone in the Miskin and the Taffs Well regions indicate relatively deeper waters. On the other hand,

---

[12]The term "partial indicates that the Main Limestone Sea was relatively shallow throughout its history of deposition. That is, it was neither emerged nor submerged completely.

the alternate deposition of oolitic limestone and dolostone, for example in the Clydach region indicates relatively shallow waters.

The common alternate deposition of the crinoidal limestone and the clay-rich limestone in the Lower Main Limestone rocks in Miskin and the Taffs Well region. During the temporary "transgressive phase" the crinoids would set up their "screen-in effect" mechanism in the manner described by Soderman and Carozzi (1962). As a result of this, the crinoids favored further growth of their own as well as of other related organisms (for example, bryozoa, brachiopods, corals, etc.) in the early part of the Main Limestone time; thus the crinoidal limestone was formed. The formation of such crinoidal limestone was subsequently interrupted by the influx of the detrital material, fine or coarse, became sufficiently high to cause the destruction of the "crinoidal screening" consequently the clay-rich crinoidal limestone, or the so-called argillaceous limestone, was formed.

The alternate deposition of the oolitic limestone and the dolostone in the Clydach region could have formed under a critical control, exerted by a barrier (possibly a reef barrier). Thus during the effective barrier time, the restricted sea would mark a low energy environmental site of deposition time, the restricted sea would mark a low energy environmental site of deposition, consequently permitting the formation of dolostone (or the dolomitized limestone). However, when the barrier was broken due to the rise in the sea level, the resulting high energy environment would favor the formation of oolitic limestone.

EVIDENCE FOR STAGE A: Although the evidence for the moderate instability of the area of deposition stems from the work of George (1956b, 1958) and Owen (1954) and others. Field geological, petrographic, and chemical observations support recurring instability in the Main Limestone Series as observed from the alternate deposition of the limestone and dolostone in the Taffs Well suggesting depth changes in the sea, presumably in response to tectonism (Pl. Ib).

The relatively open and deeper nature of the sea is shown by these observations: (1) relatively greater numbers of marine fossils found from this part of the Series (Lower ZC_1 zone, Table I), for example, more abundantly in the Miskin area, less abundantly in the Taffs Well region, and the least in the Clydach region, however, with the exception of some local areas, (2) relatively less dolomitized nature of the lithology, that is, still the lime facies dominates especially in Miskin and to a less extent the Taffs Well region, and (3) common occurrence of the microfacies I-biosparite (Pl. IVa) in which the allochems of the crinoids, bryozoa, brachiopods, and the related organisms are cemented by a "sparry calcite", suggesting a high energy of deposition (Folk, 1959).

Stage B: Partial Emergence Phase

With the passage of time, the deep and the open sea conditions of Stage A assumed the characteristics of a shallow and semi-restricted sea which later became more of a lagoon phase (Fig. 10). The subsidence-sedimentation relation during the time became geographically variable; for example, these were well-balanced in the Taffs Well region where the Main Limestone rocks are characterized by a thick and relatively uniform lithological character; whereas, in the Clydach region, their unbalanced relation resulted in the formation of irregular thicknesses and the varying lithological characters of the corresponding rocks.

Considering the nearshore location of the Clydach region (George, 1954, 1958), the differential topography (or more appropriately the hydrography) which presumably resulted from the unbalanced subsidence-sedimentation relation, must have been also further complicated by the influx of the terrigenous material, current and wave action, and by the frequent sea level changes and so forth. Although insufficient evidence is available, it is assumed that the general physiographic make-up of the Clydach region (and north of it) may well have been quite complex and similar to that observed in the Persian Gulf (Evans *et al*, 1964; Kinsman, 1964). Moreover, with regard to the Clydach region, the differential hydrography must have given rise to local depressions and swells resulting in the pockets of varying salinities, light, substrate etc., which in turn determined the faunal assumblages. It is believed that one of the features of this complex region was the development of the reffs built by the calcareous algae and corals. The algae in certain cases would have built up the calcareous pads in the hypersaline water (Carozzi, 1962) thus guaranteeing the local reef growth (see chapter 8 for more information).

By analogy with the Recent sediments, it is postulated that supratidal or *sebkha* type of environmental conditions may well have prevailed farther north of the Clydach region (Fig. 10).

This contention is supported by the occurence of evaporite deposits in these rocks (Bhatt, 1975).

EVIDENCE FOR STAGE B: The remarkably uniform lithological characters including the coarse-medium grain size, crystalline, hard and compact dolostone in much of the area of the Taffs Well region with a considerable thickness (over 1000 feet in the Tongwylais) unquestionably reflects sustainment of uniform conditions of deposition for a long period of time. Such persistently uniform conditions can be achieved only from a well-balanced subsidence-sedimentation relationship. In fact, the very striking uniformity of the lithology of the Stage B in the Taffs Well region impressed earlier workers (Dixon, 1907; Dixey and Sibly, 1918). The greater stability of the area of deposition as maintained by the well-balanced subsidence-sedimentation relation is also shown by the relatively less frequent major alterations of the limestone and dolostone (Pl. Ib).

Further support for the uniform lithology reflecting the well-balanced subsedence-sedimentation relation in the Taffs Well region emerges from the (1) commonly occurring (M.F. II dolobiomicrite (Pl. Vb) in which the dolomite idiotopes or the hypiotopes of uniform size 100-500 microns are so common. (2) uniform Ca/Mg ratio behavior which falls consistently below the value of 2:1 (Bhatt, 1972, 1973) (3) the magnesium content consistently remaining between 40-50 mole % $MgCO_3$ (Bhatt, 1973).

Farther north and northeast of the Taffs Well region, in contrast to the uniform lithologic, petrographic, and chemical properties of the Main Limestone rocks (of C_2S_1 zone) in the Taffs Well region, reflecting a well-balanced sedimentation-subsidence relation, the depositional history of corresponding rocks in the Clydach region, is revealed by (1) the patchy dolomitization (see also George, 1954, 1956a), (2) presence of the local coral beds forming thin lenses, (3) the varying lithologic properties including the variation in grain size and the quartz content, and (4) the varying chemical properties, for example, the Ca/Mg ratio behavior in vertical section at Black Rock varies considerably in relation to the Taffs Well region (Bhatt, 1976).

Stage C: Partial Resubmergence Phase

This stage marks the return of the Sea in the latter part of the Series under study (*i.e.* the S_2 zone), consequently reestablishing the high energy environmental conditions of deposition and favoring the large scale oolitic limestone deposition. However, in the Taffs Well region, dolomitization of the original carbonate material still continued but with decreasing intensity in the basal S_2 zone part of the Series. Later, this area too gave way to the deposition of the oolitic Limestone, thus signifying a shift from a semi-restricted shallow water and low energy environment to a relatively open and high energy environment of deposition during the *Stage C*.

EVIDENCE FOR STAGE C: The change in the area of deposition from a semi-restricted sea into an open and wide sea during Stage C is shown by these observations: (1) the field observations indicate the presence of a relatively less dolomitized to almost unaltered limestones in the upper part of the Series (i.e., in the S_2 zone), (2) similarly, the regional scale occurrence of the microfacies V-oosparite in the S_2 zone beds, (3) higher Ca/mg ratio or 50:1, (4) corresponding low Mg-content of less than 5 mole % $MgCO_3$ (Table IV), and (5) increase in the marine fossils-corals, crinoids, bryozoa, for example, in Miskin. Finally, Stage C in general marks the end of the major dolomitization of the Main Limestone Series.

Summary

Based upon the information at hand, the history of deposition of the Main Limestonw Series evolving with time is described in three major stages:
- (A) Partial submergence phase (open and deep sea conditions),
- (B) Partial emergence phase (restricted and shallow sea conditions) and
- (C) Partial resubmergence phase (sea becoming deep and wide).

The bulk of the dolomitization in the Series took place during *Stage B* of the partial-emergence in which the semi-restricted sea later turned into a comples-lagoon, and where salinity, depth, and the substrates varied locally, especially in the near-shore region of the Clydach.

The data at hand show that laterally the Miskin area and the Taffs Well region marked the relatively deeper water zone (possibly the intertidal zone) and the Clydach region occupied the supratidal zone with a postulated sebkha type zone existing north of this region.

Chapter Eight
DOLOMITIZATION

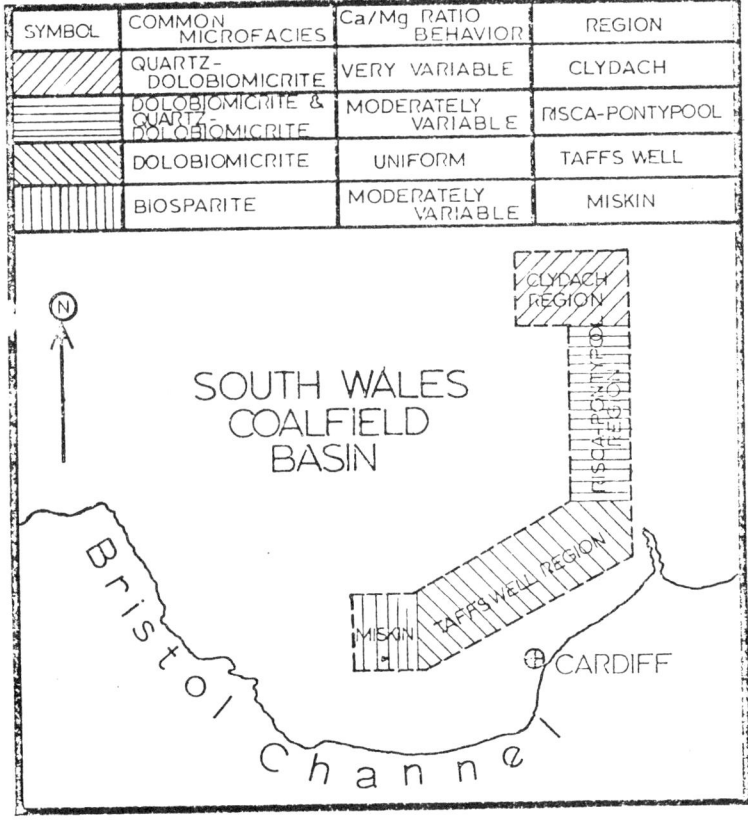

Figure 11. Lateral distribution of Ca/Mg ratios and major microfacies in the Main Limestone Series of South Wales.

CHAPTER EIGHT

ORIGIN AND MECHANISM OF DOLOMITIZATION

Introductory Remarks

Consideration of the following four governing factors is essential in comprehending the origin and mechanism of dolomitization in the Main Limestone rocks:
  (1) a favorable geologic framework of the site of deposition,
  (2) an adequate supply of magnesium,
  (3) algal growth indicative of a reef-structure, and
  (4) presence of hypersaline waters indicative of high evaporation.

Geologic Framework

The NE-SW section between the Clydach region and the Miskin area via the Taffs Well region (Fig. 1) offers but a thin remnant of the original larger outcrop of the Main Limestone rocks in South Wales. Nevertheless, this narrow outcrop (of the Main Limestone rocks) provides an ancient analog of the Recent sediments which are forming today in the tropical and subtropical zones of the world. This contention is based upon the detailed chemical, petrographic and field evidences obtained in this study. The following discussion would show how all evidences at hand permit establishing a definite geologic and geochemical pattern through the NE-SW section in the area of study.

(1) The regional distribution of the major chemical parameters including the Ca/Mg ratio and to lesser extent by the K/Na ratio and the Mn/Fe ratio indicate three broad zones (Fig. 11): (i) moderately uniform zone of the Miskin and the Taffs Well regions, (ii) moderately variable zone of the Risca-Pontypool region, and (iii) the zone of greatest variation as marked by the Clydach region.

(2) The regional distribution of the major microfacies through the NE-SW section also indicate a pattern similar to one based upon the chemical parameter (Fig. 11). In general the petrographic evidence show three major zones: (i) the dominating microfacies I of *biosparite* in the Miskin area (ii) the dominating microfacies II of *dolobiomicrite* in the Taffs Well

Plate V. Stromatolitic Limestone. Locality: Risca

region and (iii) the common microfacies III of *quartz-bearing dolobiomicrite* in the Clydach region.

(3) Major field observations including the distribution of the lithologic properties of the Main Limestone through the NE-SW section indicate three major zones: (i) moderately dolomitized limestones in the Miskin area, (ii) highly dolomitized limestones (or the dolostones) in the Taffs Well region and (iii) irregularly dolomitized limestones in the Clydach region. Also, the faunal distribution in these rocks (through this section of NE-SW) indicate close trend with those of the other results as discussed above. Therefore, it is concluded that the geologic (or the geochemical) pattern through the NE-SW section of the study area, in fact represents a water-depth pattern and, which further permits to divide the area into four lateral sequences of carbonate environment. They are (i) the subtidal to intertidal zone of the Miskin area, (ii) the intertidal to lagoonal zone of the Taffs Well region, (iii) the lagoonal to supratidal zone of the Clydach region, and (iv) the postulated *sebkha* type zone existing in north of the Clydach region (Fig. 12).

Supply of Magnesium

It is obvious from the chemical data that the supply of magnesium was abundant during the Main Limestone time since (1) the Ca/Mg ratio is less than 2:1 (Bhatt 1973) and (2) the consistent high magnesium amounts of 40-50 mole $MgCO_3$ (Table IV).

The source of magnesium is believed to be principally the calcareous algae (Chave, 1954a-b) but included the Mg-leaching from crinoids, brachiopods, formaminifera. In addition, magnesium may have been supplied by some Mg-rich clays.

By considering the favorable geologic framework of the site of deposition of these rocks under question, the increase in the Mg/Ca ratio in the solution of the Main Limestone Sea must have been aided by the *refluxion* (Adams and Rhodes, 1960) and the *capillary concentration* (Illing, 1959).

Algal Occurrence

Earlier, George (1954, 1958) and Wood (1941) have already discussed the significance of algae ("algal dust" of Wood, 1941) in the formation of the calcity-mudstone rocks (belonging to the *Stage B*) in the Clydach region.

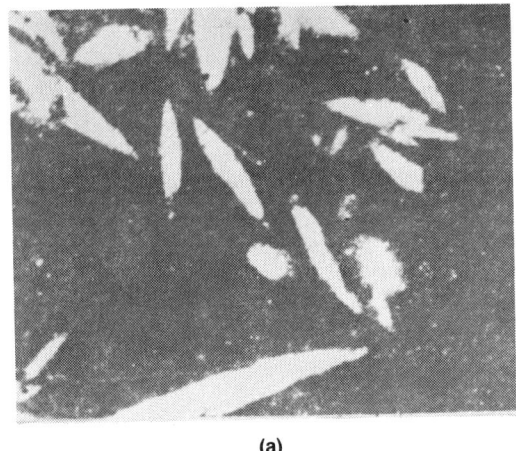

(a)

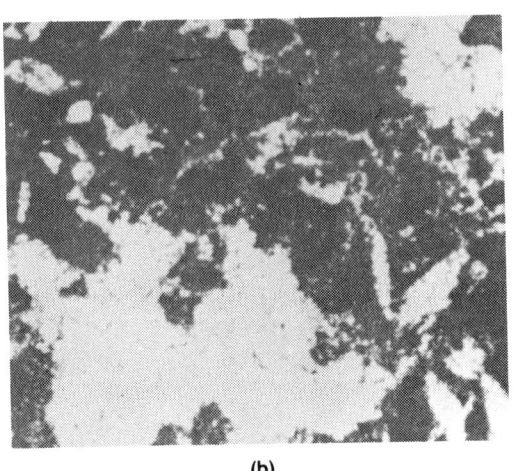

(b)

Plate VI. Dolomite pseudomorphs after an early diagenetic gypsum lenses in fine matrix as seen in the Main Limestones from the Clydach region. Nicols not crossed X 16. (After Bhatt, 1975).

Evidences favoring the presence of the algal reef structure in the Main Limestone Series during the *Stage B* are based upon these facts:

Common occurrence of the algal structures in the Main Limestones throughout the area of investigation (Table V). The algal structures include the broken tips of *Coralline* algae, the Red algae and the rings of *lithothamnium* (Table II). Moreover, discovery of *stromatolitic* limestones from Risca and the Clydach region furthersupport the sediment-binding activity of algal mats (Plate V).

Despite the fact the Main Limestone rocks are so intensely dolomitized in the study area, the algal structures have managed to survive on a regional scale (Table II). This suggests well developed growth of their reef-structure, perhaps on a larger scale than has hitherto been realized. Additional support for the presence of the reef structures in the rocks of the similar age and lithology come from their discovery in northern England and southern Scotland (Anderson, 1950).

It must be, however, pointed out that the limitation of the Main Limestone outcrop in the field (Fig. 1) and their high degree of dolomitization, preclude further attempt to establish a definite pattern of reef-structure. Therefore, the term "patch-reef" is preferred in this book.

Hypersaline Condition

Evidences indicating the hypersaline condition during Stage B emerge from petrographic and chemical observations (Bhatt, 1975). The petrographic study showed that the replaced minerals of original gypsum (or anhydrite) in the Main Limestone rocks of the area of investigation commonly occur: (1) as dolomite pseudomorphs after an early diagenetic gypsum lenses in fine matrix pelsparite (Pl. VI; See also West, 1964).

The chemical evidence is indirect which is primarily based upon the value of the Mn/Fe ratio of less than 1:1 (Fig. 8). The graph of Mn *vs.* Fe (in parts per million) suggest presence of a sulfate-bearing solution in the Main Limestone Sea during Stage B (Fig. 9).

Further support in favor of the presence of evaporates in the Main Limestone rocks come from the discovery of the Lower Carboniferous evaporite deposits from Northern Ireland (Sheridan *et al*, 1961) and from Leicestershire (Llewellyn, *et al*, 1969), especially is noteworthy. Experimental studies (for

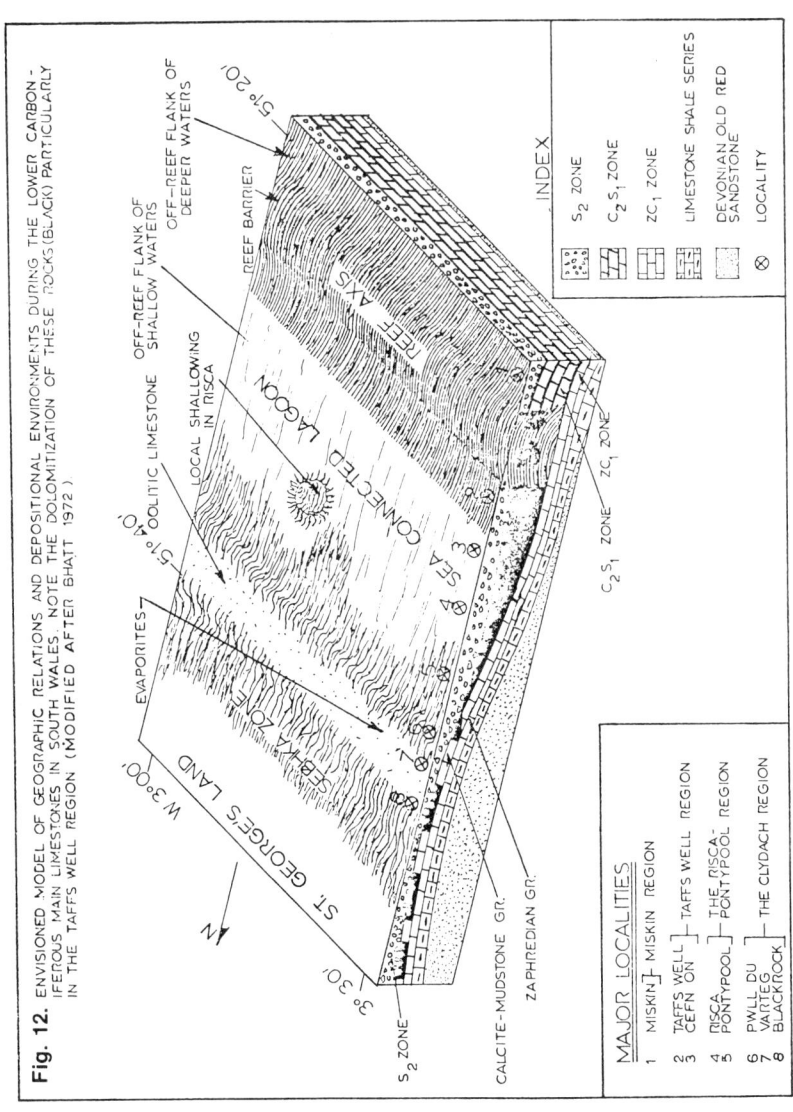

Fig. 12. ENVISIONED MODEL OF GEOGRAPHIC RELATIONS AND DEPOSITIONAL ENVIRONMENTS DURING THE LOWER CARBONIFEROUS MAIN LIMESTONES IN SOUTH WALES. NOTE THE DOLOMITIZATION OF THESE ROCKS (BLACK) PARTICULARLY IN THE TAFFS WELL REGION (MODIFIED AFTER BHATT 1972).

(Also from Bhatt, 1976)

example, Siegal, 1961; Erenberg, 1963) have already indicated the significance of the SO_4 radicals (i.e. the role of the hypersaline waters) in the process of dolomitization, and in fact are well documented by the geological studies of the Recent sediments in the Persian Gulf (Wells, 1962; Sherman, 1963; Illing et al, 1965) an in Bonnaire (Deffeys et al, 1965).

Mechanism of Dolomitization in the Main Limestone Series

By employing the four governing factors of dolomitization discussed above, a model explaining the mechanism of dolomitization in the Main Limestone rocks may be suggested (Fig. 12). According to the proposed model, the dolomitization in the Main Limestone rocks (during Stage B) was thought to have been caused by a massive refluxion of the high Mg/Ca ratio hypersaline waters which moved through the algal reef structure from the region lying north of the Clydace region and towards the deeper waters of the Taffs Well region. The downward movement or the *refluxion* of such solution was also aided by the upward movement or the *capillary concentration*, especially more effectively in the Clydach region where the waters were relatively very shallow. In response to the movement of such hypersaline solutions, the freshly deposited calcareous muds containing aragonite and high Mg-calcite (for example, in the Taffs Well and Miskin area were altered into dolomitic muds sooln after their deposition and which eventually formed the early diagenetic dolostones). The chemistry of the dolostone formation is outlined in the subsequent discussion.

Dolomitization of the rocks under study during Stage A (ZC_1 zone) must have been relatively less effective in virtue of the open and deep sea conditions (Fig. 13).

Origin of Dolostone in the Main Limestone Series

The origin of the early dagenetic dolostone in the Main Limestone rocks under study can be understood in terms of these representative chemical reactions.[13]

$$2CaCO_3 + Mg^{++} \rightleftharpoons CaMg(CO_3)_2 \quad \dots \dots \dots \dots \quad (1)$$

$$CaCO_3 + Mg^{++} + 2HCO_3 \rightleftharpoons CaMg(CO_3)_2 \, H_2CO_3 \quad \dots \dots \quad (2)$$

$$2CaCO_3 + MgSO_4 \rightleftharpoons MgCO_3 \quad \dots \dots \dots \dots \quad (3a)$$

$$CaCO_3 + MgCO_3 \rightleftharpoons CaMg(CO_3)_2 \quad \dots \dots \dots \dots \quad (3b)$$

In general, the reactions (1) and (2) could have operated during Stage A when the Sea was relatively deep and open but continued to operate until it became more restricted and shallow. On the other hand, the reactions (3a) and (3b) progressively became more active during Stage B when lagoonal conditions developed (Fig. 10). Since bulk of the dolomitization occurred during Stage B, reactions (3a) and (3b) are elaborated here as below.

FORMATION OF EARLY DIAGENETIC DOLOSTONE

When modified sea waters (i.e. the hypersaline solution in (Fig. 14) came in contact with the calcareous muds (e.g. in the Taffs Well region), the unstable magnesite may well have formed first (3a); since $MgCO_3$ being not in equilibrium with the $CaCO_3$ (eq.3a), it further reacted with more $CaCO_3$ to form stable $CaMg(CO_3)_2$ (eq.3b).

The reactions (3a and 3b) must have occurred at a higher state of evaporation possibly reaching the point of precipitation of gypsum simply because the energy required for the formation of Ca-Mg seed crystal is best accomplished in relatively more "contracted solutions" (Usdowski, 1969). Also the $SO_4^=$ complexed with Ca^{++} to yield gypsum (eq. 3a; also see discussion on the evaporite minerals in previous section). Thus, such complexing of $SO_4^=$ with Ca^{++} removes Ca^{++} and simultaneously favors Mg/Ca ratio in the residual solution which ultimately forms the early diagenetic dolostones.

FORMATION OF LATE-DIAGENETIC DOLOSTONE

With the passage of time, the depth of burial of the Main Limestone increased and the early dolomitization gave way to the late-dolomitization where pore solutions rather than the large scale movement of the modified sea waters (or the hypersaline waters) were the active medium.

During this stage of subsequent diagensis magnesium may have come from ions absorbed in the original calcite structure (e.g. the algal structures in particular) and from the Mg-rich pore solutions themselves.

In general, the late-dolomitization (in the Main Limestone rocks) must have been controlled by two major factors: (1) posthumously disintegrated Mg-rich algal dust derived from their original patch-reef structure and (2) the increasing load.

---

[13]The discussion on the chemical aspect of dolomitization is based upon the works of for example, Garrels et al (1960) and Kramer (1959).

The former provided uneven distribution of the needed magnesium whilst the latter blocked the flow of Mg-rich solution by decreasing the permeability of the then semi-consolidated sediments. The load effect must have also aided in increasing the temperature and pressure of these sediments, thus concomittantly raising the probability of their alteration from limestone into dolostone. The net result of the over-load and the supply of magnesium from the disintegrated algae would be the formation of the late-diagenetic dolostone with a characteristic development of patchy-dolomitization a feature which is so common in the Clydach region.

The syngenetic and the epigenetic dolostones in the Main Limestone Series are restricted in time and space, and therefore they are described briefly as follows:

FORMATION OF SYNGENETIC DOLOSTONE

Judging from their close association with the fine material presumbably represents the original algal structures and the Mg-rich clays, the syngenetic dolostone are regarded to have formed during the regressive phases when salinity, pH, magnesium supply and the water movements coincided favorably over periods of sufficient duration. Also by analogy with the Recent sediments, it is suggested that moderate scale deposition of the syngenetic dolostone (or the dolomite) may well have occurred in the postulated *sebkha* zone lying north of the Clydach region (Fig. 14).

FORMATION OF EPIGENETIC DOLOSTONE

After complete burial, lithification and post-depositional uplift of the Main Limestone rocks (Bhatt, 1976) the Mg-rich solutions of the latter geologic dates circulated through the fractured zones-faults, joints, and so forth, consequently giving rise to the "vein-dolomitization" i.e. the epigenetic dolostone of this study.

Evidences indicative of the syngenetic and the diagenetic (early and late) and the epigenetic dolostones in the Series under study are given in the immediately following Chapter.

Chapter Nine

CLASSIFICATION OF DOLOSTONES

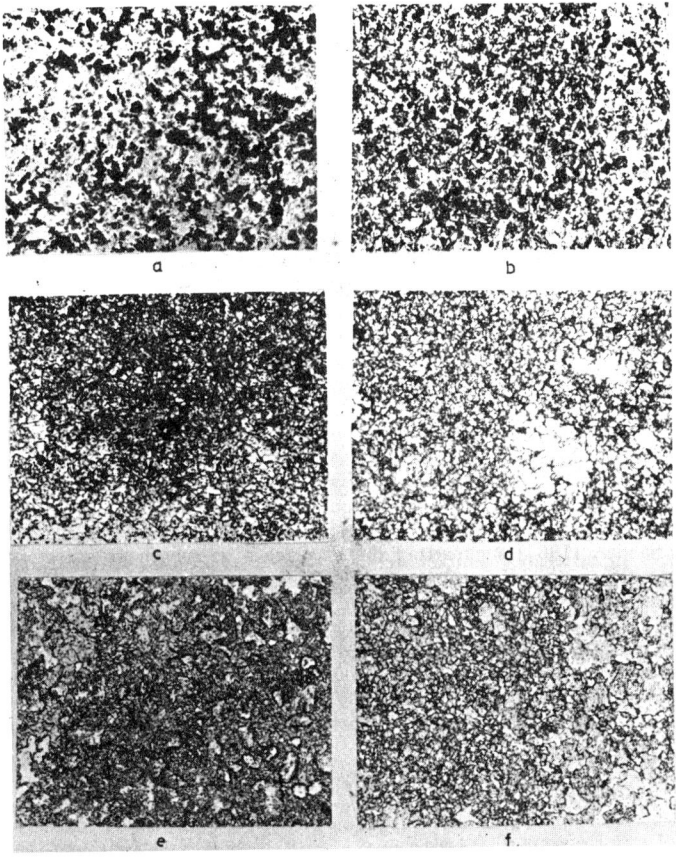

Plate VII.
Major types of dolostones in the Main Limestone. a. Clusters of perfectly idiomorphic dolomite rhombs of very small size (10 microns) appear to have developed preferentially in fine material (dark) in the intragranular space. Note the complete absence of fossil impressions. The white part represents the calcitic material. Ca/Mg ratio: 1 : 1 or less. Locality: Tongwynlais, Taffs Well region. Nicols not crossed ×96. b. Clusters of perfectly idiomorphs of dolomite having small-sized grains are seen preferentially developed within a fine material (dark). Notice the absence of fossil impressions. Some linear black spots (upper rightside) may well represent disintegrated parts of algae. Ca/Mg ratio: 1 : 1 to 1.4 : 1. Locality: Tongwynlais, Taffs Well region. Nicols not crossed ×96. c. Sparry-calcite ("orthosparite") cement filling the original organic parts (white bright) suggests an early diagenetic and authigenetic cementation. The relatively darker material characterized by the rhomb-shape grains indicate dolomite hypiotopes developing out of an original fine material (black material in the background). Locality: Tongwynlais. Nicols not crossed ×16. d. "Moldic porosity" of original organic cavities as seen filled with "drusy-calcite" cement (white-bright parts especially in the upper centre part of the photograph). The remaining portion shows the growth of dolomite in a fine grained original material (black material in the background). Nicols not crossed ×16. e. Incompletely altered oolitic limestone. Notice the black rims around the edged of oolites. Locality: Clydach region. Nicols not crossed ×16. f. Incompletely altered limestone. Notice the abrupt lateral change of grains (left to right side of the photograph). Locality: Clydach region. Nichols not crossed ×16. (Source: Bhatt, 1976)

CHAPTER NINE

TIME OF DOLOMITIZATION IN THE MAIN LIMESTONE SERIES

Introductory Remark

The dolostone occurring in the Main Limestone Series has been identified and defined as follows:

Syngenetic Dolomite

DEFINITION: Following Friedman and Sander's (1967) definition with slight modification, the syngenetic dolomite[14] in the Main Limestone Series is defined as dolomite euhedra formed penecontemporaneously in its environment of deposition as micrite or fine grained crystals, it occurs in the form of dolomite crystals disseminated in marine limestone with or without terrigenous sediments.

EVIDENCES: The geological and petrographical criteria needed for the recognition of syngenetic dolomite have been discussed by various workers (Dixon, 1907; Steidman, 1926; William, Gilbert, and Turner, 1954, p. 351; Pettijohn, 1957; Carozzi, 1960; Sarin, 1962; and Weber, 1964). Recently, Friedman and Sanders (1967) summarized the occurrence of syngenetic dolostone from the major strata of the geologic column.

The petrographic and chemical criteria favoring the syngenetic dolomite in the Main Limestone Series include: (1) The presence of almost perfect idiotopes of dolomite rhombs having uniform grain size of less than 10 microns (Pl. VIIa); such dolomite idiotopes occur (a) either in disseminated or clustered form within a fine matrix showing no signs of significant grain growth (Pl. VIIb); or (b) found closely associated with the terrigenous material within the marine limestone; for example, as observed in the clay-rich limestone. These observations are in general agreement with Friedman and Sander's (1967) definition of syngenetic dolomite.

---

[14]The term "dolomite" instead of "dolostone" is followed here since the cyrstals of dolomite euhedra indicative of syngenetic origin are observed in the rocks under study.

(2) The abundant supply of magnesium needed for direct precipitation of dolomite crystals is indicated by (a) very low Ca/Mg ratio (Bhatt, 1973) 1:1 (check samples J305 and J306) and (b) high Mg-content of more than 50 mole % $MgCO_3$ (Table 5). (3) The relatively high trace element composition and the high clay % (Table VII) which is in accordance with Weber's criteria of his primary dolostone; and (4) The absence of fossil impressions except for the suspected fine debris of algae (Pl. VIIa) implies the presence of saline waters which must have favored the precipitation of syngenetic dolomite.

CONCLUSION: The samples indicative of the syngenetic dolomite in the Main Limestone Series are few indeed and they seem to be restricted to the Taffs Well region as no rocks possessing the above mentioned properties were encountered elsewhere in the study area. By analogy with the Recent sediments, it is possible that syngenetic dolomite may have formed on a moderate to small scale in the postulated sebkha zone of the study area.

Early Diagenetic Dolostone

DEFINITION: The early diagenetic dolostone is a carbonate rock which is formed by the metasomatism of $CaCO_3$ by $MgCO_3$ in freshly deposited calcareous sediments. The reaction solution of the early diagenesis is the modified pre-concentrated water (Udowski, 1969). The term, early diagenetic dolostone, corresponds with the "penecontemporaneous dolomite" (Dixon, 1907) or to the "early secondary dolomite" of many American workers.

EVIDENCE: The petrographic and the chemical criteria suggestive of the early diagenetic dolostone in the Main Limestone Series include presence of:

(1) characteristic poorly preserved fossils like the crinoid issicles;

(2) commonly occurring phantom fossil structures, organic cavities and pore spaces filled by drusy calcite mosaic (Pl. VIId);

(3) relative coarseness of grains in the form of hypiotopes or sometimes idiotopes of dolomite rhombs (Pl. VIIc);

(4) fairly uniform Ca/Mg ratio (Table III);

(5) uniform Mg content ranging 40-50 mole % $MgCO_3$ (Table IV) and

(6) fairly low trace element composition of K, Na, Mn, and Fe and the clay % (Table VII).

Items (1) and (2) indicate the original clacareous nature of the sediemtns, whereas numbers (4) and (5) suggest a uniform environment of deposition which could develop in the early stages of diagenesis when the sediments were highly porous and consequently their effective permeability would allow a mass scale mobilization of the Mg-bearing solutions. The point (6) is in general accordance with Weber's (1964) "secondary dolostone".

CONCLUSION: The majority of the analyzed samples from the study area showing the aforementioned chemical and petrographic properties came from the Taffs Well region. Therefore, it is concluded that the early diagenetic dolostone in the Main Limestone Series predominates in the Taffs Well region.

Late Diagenetic Dolostone

DEFINITION: The late diagenetic dolostone is a carbonate rock which is formed by the metasomatism of $CaCO_3$ by $MgCO_3$ at a later stage of diagenesis when the sediments were either semi-consolidated or consolidated by the increase in debth of burial but prior to uplift and fracturing. The reaction solutions of the late diagenesis consists of pore-solutions (Udowski, 1969). The term "late diagenetic" dolostone corresponds with the "subsequent dolomite" of George (1954, 1956a).

EVIDENCE: Major evidences supporting the late-diagenetic dolostone in the Main Limestone Series are as follows:
  (1) Patchy dolomitization of these rocks, especially in the Clydach region as observed in the field (see also George, 1954).
  (2) Petrographic evidences include the heterogenous nature of these rocks, for example, (a) varying grain size, and (b) non-uniform texture characterized by different degrees of grain growth.
  (3) Different degrees of fossil replacement indicating dolomitization of different intensities perhaps occurring intermittently in the later stages of diagenesis.
  (4) Chemical evidences include (a) varying behavior of the Ca/Mg ratio, for example in the Clydach region (Table III).

Thus the very hetrogenous nature of these rocks shown

in the field, petrographically and chemically, strongly indicates a lack of uniform conditions of deposition which otherwise are expected during the very early diagenesis of the sediments. Therefore, it is believed that the Main Limestone rocks showing the variable properties of grain size, texture, partially-altered fossils, load-effect, Ca/Mg ratio, and trace elements content--are late-diagenetic in origin.

CONCLUSION: The majority of the analyzed samples in the Main Limestone Series of the study area came from the Clydach region and to a lesser extent from the Taffs Well and the Rica-Pontypool areas. Therefore, it is concluded that the late-diagenetic dolostone in the Series dominates in the general area of the Clydach region.

Epigenetic Dolostone

The epigenetic dolostone is defined as very late-diagenetic rock which is formed by localized replacement of limestone controlled by post-depositional structural elements, such as faults, fractures, and joints (Friedman and Sanders, 1967). The epigenetic dolostone corresponds with the *vein-dolomite* (Dixey and Sibly, 1918) or the "T-dolostone" (Dunbar and Rodgers, 1957).

The epigenetic-dolostone is distinguished from the late-diagenetic dolostone in three respects which also constitute the basis of its identification as follows: (1) abrupt lateral change in lithofacies, (2) restricted occurrence in terms of time and space, and (3) close association with Pb-Zn deposits.[15]

The abrupt lateral change is observed in the field; for example, where the oolitic limestone abruptly changes into dolostone, under thin section (Pl. VIIf) and chemically as shown by the Ca/Mg ratio which changes abruptly (laterally speaking) from 5:1 to 2.3:1. Similar observations were also made by Robertson (1927; see Ch. 3). The point (2) is based upon the field observations of this study and was earlier reported by Dixey and Sibly (1918). Therein, dolomites are found in the S_2 zone of the Main Limestone Series and are found locally, for example, in the Portobello quarry in the Taffs Well region.

---

[15]Not attempted in this study.

Chapter Ten
CONCLUSIONS

CHAPTER TEN

SUMMARY AND CONCLUSIONS

The salient features of the Main Limestone Series in South Wales which relate to its chemical and petrographic properties, dolomitization process, and classification are summarized as follows:

1. The field data indicate that the degree of dolomitization first appears near the Miskin area increases considerably both vertically (*i.e.* in time) and laterally towards the Taffs Well region in the east and becomes increasingly complex in the Clydach region.

2. The Main Limestone Series contains complex suites of dolostones, namely; syngenetic, early diagenetic, late diagenetic and epigenetic. Both early and late diagenetic dolostones predominate the area of investigation. The early diagenetic dolostone is more common in the Taffs Well region, whereas the late diagenetic dolostone dominates in the Clydach region. On the other hand, both syngenetic and epigenetic dolostones are restricted in time and space, and are far less common than the diagenetic type.

3. A model explaining the origin and mechanisms of dolomitization in the Series under study is suggested. According to it, the early diagenetic dolostone was formed when the freshly deposited calcareous lime-muds came in contact with the high Mg/Ca ratio hypersaline brines flowing gravitationally from the nearshore zone of the Main Limestone Sea (e.g. north of the Clydach region) whereas the late-diagenetic dolostone was originated later in the history of diagenesis of these sediments when the depth of burial was moderate and thus causing the temperature and pressure to rise, which in turn raised the probability of altering the unaltered or the partially altered limestone into dolostones.

   The subordinate varieties of syngenetic dolostone were formed when the coincidence of high temperature, salinity, algal growth, Mg-rich clays and the high Mg-Ca ratio brine movements took place in a temporary "regressive" phase of sufficient duration. The epigenetic dolostone was formed by the action of Mg-bearing secondary solutions moving through the fractured rock zones.

4. Finally to understand the problem of dolomitization in the Main Limestone rocks on a larger geographical scale, studies similar to that undertaken here are recommended for the neighboring areas of Gower and Forest of Dean where this Series outcrops again.

REFERENCES

REFERENCES

Adams   J. E. and Rhodes, M. L. 1960, Dolomitization by refluxion: *Am. Assoc. Petrol. Geologists Bull.*, v. 44, pp 1912-1920.

Alderman   A. R. and Skinner, H. C. W. 1957, Dolomite sedimentation in the southeast of South Australia: *Am. J. Sci*; V. 255, pp 561-567.

Alderman   A. R., and Von der Borch, C. C. 1960, Occurrence of hydromagnesite in South Australia; *Nature*, v. 188, p. 931.

Anderson   F. W. 1950, Some reef building calcareous Algae from the Carboniferous rocks of northern England and Southern Scotland; *Proc. Yorkshire Geol. Soc.*, v. 28, pp 5-27.

Anderson   J. G. C., 1960, Geology of Cardiff region: *University Wales Press, Cardiff, Wales, U. K.* p. 44.

Anderson   J. G. C. 1971, Research on bulk minerals in South Wales: in "Mineral exploitation and economic geology." Anderson et al (editors), *Proc. University Wales Intercoll. Colloquim, Cardiff.* pp 57-58.

Anderson   J. G. C. and Owen, T. R., 1967, Structural geology of the British Isles, Pergamon Press, London.

Bathurst   R. G. C. 1958, Diagenetic fabrics in some British Diantian limestone: *Liverpool and Manchester Geol. Jour.* v. 2, pp 11-36.

Bathurst   R. G. C. 1959, Diagenesis in Mississippian calcilutites and pseudobreccia: *J. Sediment Petrol,* v. 29, pp 365-376.

Berner   R. A., 1965, Dolomitization of the mid-Pacific atolls: *Science* v. 147, pp 1297-1299.

Berner   R. A., 1966, Diagenesis of carbonate sediments: interaction of magnesium in sea water with mineral grains: *Science* v. 153, pp 188-191.

Bhatt   J. J., 1968, Microbial action on $CaCO_3$-SiO_2-H_2O system- a preliminary report, *Geol. Soc. America, Ann. Meet.,* Mexico City, Mexico, p. 26.

Bhatt   J. J., 1971. Some preliminary observations on Ca/Mg ratio behavior in Lower Carboniferous Limestone of South Wales. In: J. G. C. Anderson et al. (Editors), Mineral Exploitation and Economic Geology. *Proc. Univ. Wales Intercoll. Colloquium, Cardiff,* p. 59.

Bhatt   J. J. 1972. *Dolomitization of the Carboniferous Main limestone Series in South Wales,* Ph.D. Thesis, Univ. Wales, Cardiff, 234 p. (unpublished).

Bhatt   J. J., 1973. Ca/Mg ratio classification of Main Limestones (Mississippian) in South Wales, U. K. *Sediment. Geol.,* 10: 225-231.

Bhatt J. J., 1974. Ti/Al ratio as geochemical index of paleoenvironment — a note. *Chem. Geol.,* 13: 75—78.

Bhatt J. J., 1975. Evidence of evaporite deposition in the Lower Carboniferous Main Limestone Series of South Wales, U. K. *Sediment. Geol.,* 13: 65—70.

Bhatt J. J., 1976. Geochemistry and petrology of the Main Limestone Series (Lower Carboniferous), South Wales, U. K. *Sediment. Geol.,* 15: 55-86.

Bhatt J. J., 1956(b), Carbonate peteology of an Upper Ordovician-Silurian section at the Lone Mountain, Eureka County, Nevada. *Sediment. Geol.,* v. 15, pp. 172-191.

Bhatt J. J.. Diagenetic pattern in the Main Limestone Series (Lower Carboniferous), South Wales, U. K. *Unpubl. ms.* 20p.

Chave K. E., 1952, A solid solution between calcite and dolomite *J. Geol.,* v.60, pp 190—192.

Chave K. E., 1954(a) Aspects of biogeochemistry of magnesium-calcareous marine organisms: *J. Geol.,* v.62, pp 266—283.

Chave K. E., 1954(b) Aspects of biogeochemistry of magnesium-calcareous sediments and rocks: *J. Geol.* v62, pp 587—599.

Chester R., 1965, Elemental geochemistry of marine sediments *in chemical oceanography'* Skinner and Riley (editors) v.2, pp 23—80.

Chilingar G. V., 1956. Relationship between Ca/Mg ratio and geologic age: *Am. Assoc. Petrol Geologists Bull.,* v.40, pp 2256—2266.

Chilingar G. V., 1962, Possible loss of magnesium from fossils to surrounding environment: *J. Sed. Petrol,* v.32, pp 132—139.

Chilingar G. V., and Bissell, H. J., 1963(a), Is dolomite formation favoured by high or low pH? *Sedimentology,* v.2, pp 171—172.

Chilingar G. V., and Bissell, H. J. 1963(b), Formation of dolomite in sulfate solutions. *J. Sediment, Petrol.* v.33, pp 801—803.

Clayton R. N., and Epstein, S. 1958. The relationship between C / O ratio in co-existing quartz, carbonates and iron oxides from various geological deposits, *J. Geol.,* v.66, pp 352—272.

Cloud P. E., 1962, Environment of calcium carbonate deposition west of Andros, Bahamas: *U. S. Geol. Surv. Profess.,* Papers, v.350. 138 p.

Cox A. H., 1920, The geology of the Cardiff district. *Proc. Geol. Assoc.,* v.31, pp 45—75.

Curtis	Robin, Evans, Graham, Kinsman, D. J. J. and Shearman, D. J. 1963, Association of dolomite and anhydrite in Recent sediment of the Persian Gulf: *Nature,* v.197, pp 679—680.
Daetwyler	C. C., and Kidwell, A. L., 1959, The Gulf of Bataboo, a modern carbonate basin: *World Petrol. Congr.*, Proc. 5th, N. Y., 1959, sect. 1, p. 121.
Daly	R. A. 1907, The limeless ocean of Pre-Cambrian time. *Am. Jour. Sciences,* v.23, pp 93—115.
Daly	R. A., 1909, First calcareous fossils and evolution of limestone *Geol. Soc. American Bull.*, v.20, pp 153—170.
Daughtry	A. C., Perry, D. and William M., 1962, Magnesium isotope distribution in dolomite. *Geochim. Cosmochim. Acta,* v.26, pp. 857—866.
Deffeyes	K. S. Lucia, F. J. and Weyl, P. K. 1965, Dolomitization of Recent and Plio-Pleistocene sediments by marine evapoite waters on Bonaire, Netherlands Antiles: *in Dolomitization and limestone diagenesis - a symposium,* Pray L. C. and Murray R. C. (ed.), S.E.P.M. special Publ. No. 13, pp 71—88.
Degens	E. T., and Epstein, S., 1964. Oxygen and carbon isotope ratios in co-existing calcites and dolomites from recent and ancient sediments. *Geochim. et Cosmochim. Acta.* v.28, pp 23—44.
Degens	E. T., 1965, Geochemistry of sediments — a brief survey. Prentice-Hall, Englewood Cliffs. New Jersey, 342 p.
Dickson	J. A. 1965, A modified staining technique for carbonates in thin section. *Nature,* v.205, p 587.
Dickson	J. A. 1966, Carbonate identification and genesis as revealed by staining. *J. Sediment. Petrol.*, v. pp 491—505.
Dixey	F. and Sibley, T. F., 1918. The Carboniferous limestone series of the southeast margin of the South Wales Coalfield *Quart. Jour. Geol. Soc. London.* v.73 , pp 11—164.
Dixon	E. E. L. 1907. Notes in The geology of the South Wales Coalfield Pt. VIII. The country around Swansea by Strahan, *A. Mem. Geol. Surv., England and Wales* pp 13—40.
Dixon	E. E. L., and Vaughan A. 1911. The Carboniferous succession in Gower (Glamorganshire) with notes on its fauna and conditions of deposition. *Quart. Jour. Geol. Soc.*, v.1XVIII pp 477—571.
Eardly	A. J., 1938. Sediments of Great Salt Lake, Utah. *Bull. Am. Assoc. Petrol Geologists.* v 22, pp 1305—1411.
Evamy	E. D., 1967. Dedolomitization and the development of rhombohedral pores in Limestone: *J. Sediment Petrol.*, v.37, pp 1204—1215.

Evans	W. D. and Cox, A. H., 1956. An Old Red sandstone - Carboniferous Limestone series junction at Tongwynlais north of Cardiff. *Geol. Mag.* v 93, pp 431—434.
Evans	G. Kinsman, D. J. J. and Shearman, D. J., 1964. A reconnaissance survey of the environment of Recent carbonate sedimentation along the Trucial coast, Persian Gulf, in *'Delta and shallow marine depostis'* Van Straaten (ed.) Vol1. Elsevier publd. Amsterdam, pp 129—135.
Fairbridge	R. W. 1957. The dolomite question in *'Regional aspects of carbonate deposition, a symposium,'* LeBlanc, R. J. and Breeding, J. G. (editors), Soc. Econ. Paleontologists Mineralogists, spec. publd. 5, pp 125—178.
Fairbridge	R. W. 1964, The importance of limestone and its Ca/Mg content ot paleoclimatology in Nairn A.E.M. (editor). *Problems in paleoclimatology,* Wiley, New York, pp 431—530.
Fairbridge	R. W., 1967(a). Phases of diagenesis and authegenesis: *in Carbonate rocks,* Chilingar G. V., and Larsen (editors) Elsevier, Amsterdam, pp 19—91.
Folk	R. L., 1959, Practical petrographic classification of limestones *Bull. Am. Assoc. Petrol. Geologists,* v.43, pp 1—38.
Folk	R. L., 1965, Some aspects of recrystallization in ancient limestones. *In Dolomitization and limestone diagenesis* (Pray, L. C. & Murray, R. D. ed.), S.E.P.M. spec. Publ., v.13, pp 14—18.
Friedman	G. M. 1964, Early diagenesis and lithification in carbonate sdeiments: *J. Sediment Petrol,* v.34, pp 777—813.
Friedman	G. M., 1965. Terminology of crystallization textures and fabrics in sedimentary rocks: *J. Sediment Petrol.,* v.35, pp 643—655.
Friedman	G. M. and Sanders, J. E., 1967. Origin and occurrence of dolostones *in Carbonate rocks,* Chilingar, G. V., Bissell, H. J., and Fairbridge, R. W. (editors) Elsevier, Amsterdam, v.9A, pp 267—348.
Friedman	G. M. 1969, Trace elements as possible enviromental indicators in carbonate sediments *in Depositional* environments in carbonate rocks, Friedman G. M. (editor), S.E.P.M. sepcial pub. No. 14, pp 192—198.
Garrels	R. M., Thompson, M. E. and Siever, R., 1960, Stability of some carbonates at 25°C and one atmosphere total pressure: *Am. J. Sci.* v.258, pp 402—418.
Galle	O. K. and Runnells, R. T. 1960. Determination of Co_2 in carbonate rocks by loss on ignition: *J. Sediment. Petrol* v.30, pp 613—618.
George	T. N. 1954, Pre-Seminula Main Limestone of the Avonian Series in Breconshire: *Quartz. Jour. Soc. London.* v110, pp 283—322.

George   T. N. 1956(a), Carboniferous Main Limestone of the east crop in South Wales: *Quart. Jour. Soc. London.* v.111, pp 309-322.

George   T. N. 1956(b), The Numarian Usk anticline. *Proc. Geol. Assoc.* v.66, pp 297—316.

George   T. N. 1958, Lower Carboniferous paleogeography of the British Isles: *Proc. Yorkshire Geol. Soc.* v.31, pp 227—318.

George   T. N. 1970, *British Regional geology* — South Wales, H. M. Stationary Office, London, 152 p.

Ginsburg   R. N. 1957. Early diagenesis and lithification of shallow-water carbonate sediments in South Florida: *in Regional aspects of carbonate deposition.* Soc. Econ. Paleontologists Mineralogists, spec. publ. 5, pp 80—100.

Goldsberg   E. D. and Arrhenius G. O. S. 1958. Chemistry of Pacific pelagic sediments: *Geochim. et Cosmochim Acta.* v 13. pp 153—212.

Goldsmith   J. R., and Graf, D. L. 1958. Structural and compositional variations in some natural dolomites: *J. Sediment Petrol* v 66, pp 678—693.

Graf   D. L., and Goldsmith, J. R. 1956. Some hydrothermal synthesis of dolomite and proto-dolomite: *J. Sediment Petrol.* v. 64. pp 173—186.

Graf   D. L. (1960a), Geochemistry of carbonate sediments and sedimentary carbonate rocks Pt. 1. Carbonate mineralogy carbonate sediments: *Illinois State Geol. Surv., Circ.* 297. 39 p.

Graf   R. E., 1958. Concept of diagenesis in argillaceous sediments: *Bull. Am. Assoc. Petrol. Geologists,* v 42, pp 246—253.

Hill   W. E., and others, 161. Methods of chemical analysis for carbonate and silicate rocks: *Kansas State Geol. Surv. Bull.* 152, p. 30.

Hsu   J., 1967. Geochemistry of dolomite formation: in *Carbonate rocks,* Chilingar, G. V., and Bissell, H. J. (editors) v. Elsevier, Amsterdam, pp 169—191.

Hsu   J., and Siegenthaler, 1969, Preliminary experiments on hydrodynamic movement induced by evaporation and their bearing in the dolomite problem: Lithification of carbonate sediments, Fauchbouer (editor), *Sedimentology,* v.12, pp 11—25.

Illing   L. V., 1959, Deposition and diagenesis of some upper Paleozoic carbonate sediments in western Canada: *World Petrol. Congr. Proc.,* 5th, N. Y. 1959, v.7.

Illing   L. V., and Wells, A. J. 1964. Penecontemporenous dolomite in the Persian Gulf: *Bull Am. Assoc. Petrol Geologists,* v 48, p.532 (abstr.)

Illing   L. V., Wells, A. J. and Taylor, J. C. M., 1965: Penecontemporary dolomite in the Persian Gulf: In *Dolomitization and limestone diagenesis* (Pray, L. C., and Murray, R. C. (ed.) Soc. Econ. Paleontologists Mineralogists, sp. pub. No. 13, pp 89—112.

Kahle   C. F., 1965. Possible role of clay minerals in the formation of dolomite: J. Sediment Petrol., v.35. pp 448—453.

Katz   A., and Friedman, G. M. 1965. The preparation of stained acetate peels for the study of carbonate rocks: *J. Sediment Petrol.* v.35, pp 248—249.

Kinsman   D. J. J., 1964. The Recent Carbonate sediments near Halat el Bahrani, Trucial coast, Persian Gulf. in 'Delta and shallow marine deposits' Von Straaten ed. Vol. 1, Elsevier pub. Amsterdam, pp 129—135.

Kramer   J. R., 1959, Correction for some earlier data on clacite and dolomite in sea water. *J. Sediment Petrol.*, v.29 pp 456—69.

Krauskopf   K. B., 1957. Separation of managese from iron in sediment processes: *Geochim. Cosmochim. Acta.* v.12, pp61—84.

Krumbein   W. C. 1947. Analysis of sedimentation and diagenesis: Bull. *Am. Assoc. Petrol Geologists.* v 31, pp 168—174.

Kutznestov   S. I. Ivanov, M. V. and Lyalikova, N. N. 1964, *Introduction to geological microbiology.* McGraw Hill, New York, N. Y. 252 p.

Lalou   C. 1957, Studies of bacterial precipitation of carbonate in sea water. *J. Sediment Petrol.* v 27, pp 190—195.

Llewllyn   P. G. Backhouse, J. and Hoskin, I. R. 1969. Lower-middle Tournasian miospores from the Hathern anhydrite series, Carboniferous limestone Leicestershire, *Proc. Geol. Soc. London,* No. 1655, pp 85—91.

Lowenstam   H. A. and Epstein 1957. On the origin of sedimentary aragonite needles of the Great Bahama Bank: *J. Geol.* v 65. pp 364—375.

Miller   D. M., 1961. Early diagenetic dolomite associated with salt extraction process, Inagua, Bahamas: *J. Sediment., Petrol.* v.31, pp 473—476.

Milner   H. B. 1962. *Sedimentary petrography:* George Allen and Urwin Limited, London, 715 p.

Monaghan   P. H. and Lytle, M. L. 1956, The origin of calcareous ooliths: *J. Sediment Petrol* v 26, pp 111—118.

Middleton   G. V. 1961. Evaporite solution breccias from the Mississippian of southwest Montana: *J. Sediment Petrol.* v 31, pp 189—195.

Moore — L. R., 1947. The sequence and structure of the southern portion of the east crop of the South Wales Coalfield. *Proc. South Wales. Inst. Engr.* v 60 pp 141—227.

Moorehouse — W. W. 1959. The study of rocks in thin section: Harper Geoscience Series. New York, 370 p.

Nehrer — J. and Rohrer E. 1958. Dolombildung unter mitworkung von baktereien: *Ecologae Geol.* Helv., v 51. pp 213—315.

Newell — N. E. and Rigby, J. K. 1957. Geological studies on the Great Bahama Bank *in Regional aspect of carbonate deposition,* (editor), Soc. Econ. Paleontologists Mineralogists, spec-publ. pp

Nicholls — G. D. and Loring, D. H. 1960, Some Chemical data on British Carboniferous sediments and their relationship to the clay mineralogy of these rocks: *Clay Min. BBull. Min. Soc. London,* v 4, pp 196—207.

Oppenheimer — C. H. and Master, I. M. 1963. Transition of silicate and carbonate crystal structure by photosynthesis and metabolism: *Geol. Soc. Am. Progr. 1963. Ann. meeting* p. 125 a (abstract).

Owen — T. R. 1954. The structure of the Neath disturbance between Bryniau Gleision and Glynneath; South Wales. *Quart. Jour. Geol. Soc.* v 109 pp 333—365.

Owen — T. R. 1964, The tectonic framework of Carboniferous sedimentation in South Wales: *In Developments in sedimentalogy* L. van Straten (editor) Elsevier, Amsterdam, v.1, pp 301—307.

Pettijohn — F. J. 1957. *Sedimentary rocks:* Harper and Bros. N. Y. (2nd ed.) 718 p.

Robertson — T. 1927. The geology of the South Wales Coalfield. Pt. 2 Abergavenny. *Mem. Geol. Surv. England and Wales.* H. M. Stationary Office p. 145.

Rodgers — J. 1954. Terminology of limestone and related rocks: an interim report *J. Sediment Petrol.* v. 24, pp 225—234.

Sarin — D. D. 1962. Cyclic sedimentation of primary dolomite and limestone: *J. Sediment Petrol.* v 32, pp 451—471.

Sass — E. 1965. Dolomite-calcite relationships in sea water. The retical considerations and preliminary results. *J. Sediment. Petrol.,* v-35, pp 338—348.

Schlanger — S. O., 1957. Dolomite growth in coralline algae. *J. Sediment Petrol.* v. 27. pp 181—186.

Schmidt — V. 1965, Facies, diagenesis and related reservoir 35 properties in the Gigas Beds (Upper Jurassic), Northwestern Germany in *Dolomitization and limestone diagenesis* Murray, R. C. and Pray, L. C. (editors) S.E.P.M. spec. publ. 13. p. 180.

Shearman	D. J. 1963, Recent anhydrite, gypsum, dolomite and halite from the coastal flats of the Arabian shore of the Persian Gulf: *Proc. Geol. Soc. London.* v 1607. pp 63—65.
Shearman	D. J. Twyman, J., and Zand Karimi, M., 1970. The genesis and diagenesis of oolites: *Proc. Geol. Assoc.* v 81. pp 561—575.
Sheridan	D. J. R. Hubbard, W. F. and Oldroyd, R. W. 1967. Tournasian strata in Northern Ireland. *Sci. Proc. Royal. Dublin Soc.* v3, pp 33 —37.
Shinn	E. A. Ginsburg. R. N. and Lloyd, R. M. 1965. Recent supratidal dolomite from Andros Island, Bahamas *in Dolomitization and limestone diagenesis* — a symposium, Murray R. C., and Pray L. C. (editors) S.E.P.M. Spec. publ. 13, 180 pp.
Siegal	F. R. 1961. Factors influencing the precipitation of dolomite carbonate rocks: *State. Geol. Surv. Kansas, Bull.* 152 (5). pp 127—158.
Skinner	H. C. 1963. Precipitation of calcian dolomites and the magnesian calcites in the southeast of South Australia: Am. *J. Science,* v.261. pp 449-472.
Steidman	E. 1926. The origin of dolomite *in* Twenhofel, W. H. (editor) *Treatise in sedimentation,* Williams and Wilkins, Baltimore pp 256—265.
Stehli	F. G. and Hower, J., 1961, Mineralogy and early diagenesis of carbonate sediments. *Jour. Sediment. Petrology,* v.3, pp 358—371.
Stockman	K. W. Ginsburg, R. N. and Shinn, E. A. 1957. The production of lime mud by algae in South Florida: *J. Sediment Petrol* v 37. p 633—648.
Squirrell	H. C. and others, 1969. Geology of the country around Newport, Monmouthshire) H. M. Stationary Office, p.333.
Strahan	A., 1909. The geology of the South Wales Coalfield. Pt.I The country around Newport, Monmouthsire. 2nd ed. *Mem. Geol.*
Strahan	A., 1909. On passage of a seam of coal into a seam of dolomite *Quart. Jour. Geol. Soc. London.* v 58, pp 297—306.
Taft	W. H. 1967(a). Modern carbonate sediments: *in Carbonate rocks,* Chilingar G. V. Bissell, H. J. and Fairbridge, R. H., (editors) Elsevier, Amsterdam, A. pp 29—50.
Taft	W. H. 1967(b), Physical chemistry of formation of carbonates *in* Chilingar. G. V. Bissell, H. J. Fairbridge R. H. (editors), *Carbonate rocks:* Elsevier. B. pp 151—167.
Terry	R. and Chilingar, G. V. 1955. Charts aid in studying sedimentary formations by M. S. Shevstov: *J. Sediment. Petrology* v 25. pp 229.

Udowski	H. 1968. The formation of dolomite in sediments *in Recent developments in carbonate sedimentalogy in Central Europe.* Miller, G. and Friedman, G. M. (editors), Springer — Verlag, Berlin. pp 21—32.
Van Tuyl	F. M. 1916. New points on the origin of dolomite: *Am. J. Sci.,* v 42. pp 249—266.
Vaughn	A., 1905. The paleontological sequence in the Carboniferous limestone of Bristol Area: *Quart. Jour. Soc. London.* v. 61 (242) pp 81—307.
Vogel	A. I., 1961. Text book of quantitative inorganic analysis including elementary instrumental analysis (3rd ed.) Longsmans 1216 p.
Vishnyakov	S. G., 1951, Genetic types of dolomite rocks. *Dokl. Akad. Nank S.S.S.R.,* V. 76(I), pp 112—113.
Warne	S. J. 1962, A quick field or laboratory staining scheme for differentiation of the major carbonate minerals: *J. Sediment. Petrol* v 32, pp 29—38.
Weaver	C. E. 1958. Geologic interpretation of argillaceous sediments. Part 2. Clay petrology of Upper Mississippian — Lower Pennysylvanian sediments of Central United States: *Am. Assoc. Petrol. Geologists.* v. 42. pp 272—309.
Weber	J. W. 1964, Trace element composition of dolostones and dolomites and its bearing on the dolomite problem. *Geochim. Cosmochim Acta,* v. 28, pp 1817—1868.
Welby	C. W. 1958. Occurrence of alkali metals in some Gulf of Mexico sediments: *J. Sediment. Petrol.,* v.28, pp 431—452.
West	I. M. 1964, Evaporite diagenesis in the Lower Purbeck Beds of Dorset: *Proc. York. Geol. Soc.* v 34, pp 315—330.
West	I. M. Brandon, A. and Smith, M. 1968. A tidal flat evaporitic facies in the visean of Ireland. *J. Sediment Petrol,* v 38. pp 1079—1093.
Zobell	C. E. 1957. Becteria *in 'Treatise on marine ecology and paleoecology.* 2. Paleoecology. Ladd, H. S. (editor), Geol. Soc. Am. Mem. v 67. pp 693—698.

APPENDIX

APPENDIX

LOCALITIES

A. TAFFS WELL REGION:
I. Garthwood Area:
1. A disused quarry (10928233)
2. A large quarry (11608242)
3. A quarry (11688272)
4. Main entrance to Garthwood mine (11668276)
5. Steetly quarry (12408270)
6. Cwarrw Glas quarry (12208290)
7. Barry Railway cutting (12498252)
8. Morganstown quarry (12208210)
9. Tongwynlais (along the old Cardiff Railway line) Now along the Motor Railway.
10. Exposures in small Garth Hill area.

II. Fforest Fawr and Thornhill Area:
11. Hobbs quarry (12808273)
12. Portobello quarry (12758328)
13. Tongwynlais section (former high level Barry Railway cutting, and low level Cardiff cutting (12778320)
14. Fforest Fawr
15. Nant-y-Forest Cefn-graw quarry (12758295)
16. Gelli quarry (14498324)
17. Bwlch-Cwm quarry (14508404)
18. Disused quarry (14338452)
19. Small quarry (14788460)
20. Invicinity of Cefn Carnau stream (15328450)

III. Thornhill:
21. Blaen-Nofydd quarry (16058465)
22. Small quarry (15828468)
23. A disused quarry (15918488)

IV. Cefn Onn - Rudry - Draethen:
24. Basic Slag quarry (17408515)
25. Lower quarry (17318534)
26. Cefn Onn Farm quarry (18168574)
27. Cwm Leyshon quarry (21168686)

V. Machen-Risca:
28. Machen quarry (22208870)
29. Railway cutting (22138857)
30. Road cutting
31. Risca quarry (23358980)
32. Dan-y-Craif quarry (23469090)
33. Buck Farm (23269094)
34. Railway cutting (232491913)

VI. Risca-Pontypool:
35. Ysgwbor-newdd (24289198)
36. Cwmbran Two quarries overgrown with vegetation (poor exposure).
37. Cwm Ynysicoy quarry (28209970)

B. CLYDACH REGION:

VII. Clydach Area:
38. A disused quarry, Abersychan near the river.
39. Abersychan quarry.
40. Varteg disused quarry.
41. Pwll Du quarry near Pwic-Du.
42. Quarries along the Black Road (old lime kilns)
43. Black Road quarry
44. Llanelli quarry, Gilwern Hill.
45. Tryfil quarry.

C. MISKIN REGION:

VII. Miskin Area:
46. Quarries along the motor road
47. A disused quarry (near a private farm)
48. Road cut (behind a private farm)
49. Wenvoe quarry near Barry
50. Quarries near Bridgend